Getting Started with Onshape

Fourth Edition

T0172356

Elise Moss

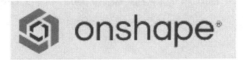

Authorized Partner

PUBLICATIONS

SDC Publications
P.O. Box 1334
Mission, KS 66222
913-262-2664
www.SDCpublications.com
Publisher: Stephen Schroff

Release Date: January, 2023

ISBN-13: 978-1-63057-576-2
ISBN-10: 1-63057-576-3

Printed and bound in the United States of America.

Preface

The book is geared towards users who have no experience in 3D modeling and very little or no experience with AutoCAD. Some experience with a computer and using the Internet is assumed.

OnShape is an exciting software tool for CAD users. Because you can use the software for FREE, that opens up CAD to anybody who is interested in creating their own models, including members of the bourgeoning Maker community and students who want to learn design tools. Because Onshape is 100% cloud-based, it is updated constantly. This means that by the time you get this book, the software will have changed. New features will have been added and the interface may look different. The good news is that the tools as outlined in this book will still work.

When I initially wrote this book, it was a good five years before the COVID-19 pandemic. When schools moved to remote learning, most schools issued their students Chromebooks. Chromebooks use a Google OS and are an inexpensive laptop. Because Onshape is 100% cloud-based and requires no software installation, many schools adopted Onshape as the CAD program to teach their students design and engineering. As a result, the demand for this textbook increased, so I am updating this textbook to meet that demand and support the continued use of Onshape.

I have provided a chapter that guides the user through how to create a model all the way to export to an stl file – which can be used to create a 3D print. Just email or take the stl file to your local Staples, Tech Shop, or other location where you can print out an stl file, and you have taken the first step to the MakerFaire journey.

I have endeavored to make this text as easy to understand and as error-free as possible...however, errors may be present. Please feel free to email me if you have any problems with any of the exercises or questions about Onshape in general. My email is elise_moss@mossdesigns.com

---

Acknowledgements

A special thanks to Mark Cheli, McKenzie Brunelle, Cody Armstrong, Noa Flaherty, Lou Gallo, Joe Dunne, Richard Doyle, Jon Hirschtick and John McEleney for their support and encouragement while writing this text. They are all enthusiastic evangelists for Onshape and are tirelessly working to improve the Onshape product so that it becomes the gold standard for CAD software.

Thanks to Stephen Schroff, Zach Werner, and other members of the SDC Publications team, who work tirelessly to bring these manuscripts to you, the user, and provide the important moral support authors need.

My eternal gratitude to my life partner, Ari, my biggest cheerleader throughout our years together.

Elise Moss
Elise_moss@mossdesigns.com

Contents

Chapter 8: Drawings 8-1

Chapter 9: Stop Base Project 9-1

Chapter 10: Pulley Project 10-1

Appendix: Onshape's App Store A-1
About the Author

Chapter 1: Getting Started

Explanation of how Onshape Works

When I started working in CAD in 1982, the emerging software – the new kid on the block – was AutoCAD created by a company headquartered in San Rafael, California, named Autodesk. The interface consisted of a screen menu. You used a keyboard and short cut keys – like L for Line – to create your technical drawings, which consisted of lines, arcs, and circles. There was no rendering. There were no layers. There was no color. You drew in white on a black background. At the time, many drafters and designers said that CADD (Computer Aided Drafting/Design) software would never replace pencil and paper.

So, here we are in 2023 and how the world has changed. Today most CADD software is moving to the cloud. This means that users will download a small software program that allows them to log in to an account and save their work on remote servers. Users are no longer tied down to a workstation or an office. You can access your work from any device anywhere in the world – as long as you have an internet connection.

Onshape takes it one step beyond this.

Firstly, Onshape is entirely 100% cloud based. This means you don't have to install any software. You don't have to get permission from your IT department to put the software on your workstation, and you don't use up any space on your hard drive or jump drive or wherever you like to store files. Everything happens on Onshape's servers.

Because Onshape is entirely cloud-based, it is accessible from any device that uses a browser, like Firefox or Chrome. This means that you can pull up Onshape with any internet connection on your tablet, smart phone, laptop, or workstation. Onshape even offers apps for Android and iOS devices.

Secondly, Onshape offers a *free* version to users. The free version is perfect for students and hobbyists. You can open an account without a credit card using an email address. The free account limits the amount of space you are provided, but you can always delete a project to free up space and start over.

Thirdly, because Onshape is entirely cloud-based, you are always using the latest version of Onshape. Onshape is regularly updated to fix bugs and add new features. Because Onshape is constantly being updated some of the screenshots you see in this text may look different from what you see on your screen. Try to ignore any differences. The tools should work the same way even if new features have been added.

In 2020, when the COVID-19 pandemic hit, many schools were forced to turn to distance or remote learning. Students were issued Chromebooks because they are cheap and easy to use. Onshape – being entirely browser based – became the software of choice for engineering and design students.

Onshape works best using Chrome or Firefox browsers. It will not work at all with Internet Explorer or Bing. Onshape requires WebGL, which IE does not currently support.

Onshape currently supports these tested and approved browsers:

- Google Chrome

- Mozilla Firefox

- Safari (Mac OS only)

- Opera

- Microsoft Edge

Onshape does use "cookies," so make sure you set your browser to accept cookies. If the browser interface is not working properly, perform a hard refresh. (Hold down Command and Shift, then press R in any Apple or Mac operating system. For a WINDOWS operating system, press Ctl+F5.)

Onshape offers an Education Enterprise subscription which can be used for programs with more than 50 students. The Enterprise program provides greater ease of setting up the software for use by multiple students. It doesn't require that each student set up their own account. Instead, the instructor or IT department can create accounts for registered students or create a single sign-on, such as LaneyCollegeStudent with a password, so more than one student can sign into the account at a time. Instructors can create groups or teams to work on a project. Instructors have access to analytics, providing insight into how much time a student is spending using the software as well as how they are using the software. This can be especially useful with remote learning to monitor how engaged each student is in the assigned work. Instructors can also assign tasks and create an engineering release process. This might be useful when working in a Fab Lab or machine shop. Students can create designs to be 3D printed or machined and then submit those designs for approval and released to be machined or printed.

Whether students are working on an individual educational subscription or an enterprise subscription, the actual tools within Onshape used to create designs and drawings are identical.

Onshape does host regular design challenge contests for students and provides support for First Robotics and Vex. If you are an educator with a student team competing in a design challenge using Onshape, reach out to Onshape for stickers and other goodies for your team.

Setting Up an Account

Estimated Time: 10 minutes

Objectives:

- Create an Onshape user account

To get started, go to onshape.com.

1. Click **Request a Trial**.

2.

    Students should sign up for the EDU plan.

    Makers or hobbyists should sign up for the free public plan.

    The free public plan means that all your files are public and available for anybody to use. If you want to secure your files, you will have to pay to use Onshape.

    Click on **Visit the EDU Plan**.

3.

    Click on **CREATE EDU ACCOUNT**.

4.

Sign up for Onshape for Students and Educators

First Name *

ELISE

Last Name *

MOSS

Email *

elise_moss@mossdesigns.com

Are you a student or educator? *

Educator

School level *

College or University

☑ * I acknowledge that my personal details will be processed in accordance with Onshape's Privacy Policy [↗]. PTC will not use student email addresses for any marketing purposes.

CREATE EDU ACCOUNT

- Affiliates: Onshape may share your personal information with PTC Inc's Affiliates where it is necessary to support the provision of the Services or it is necessary for the Affiliates to process the personal information to support their legitimate business interests. All of PTC Inc's Affiliates are required to honor this Privacy Policy.
- Authorized Resellers and Partners: We may also share your personal information with our authorized resellers and partners for the purpose of further supporting your use of Onshape, and for co-marketing activities. All such authorized resellers and partners are required by Onshape to provide you with a means of opting out of some or all of their communications with you.
- Aggregated and Non-Identifiable: Onshape may share aggregated or deidentified personal information with our partners or others in our sole discretion, including for business or research purposes.

Fill in the dialog box.

Click **CREATE EDU ACCOUNT**.

Note: *You have to initially opt in to receive spam, but you can opt out afterwards.*

It is up to each individual user to manage their opt-out settings every time you receive an email from an Onshape partner.

This may or may not be a concern, but you should be aware.

Students who are minors are protected from receiving spam.

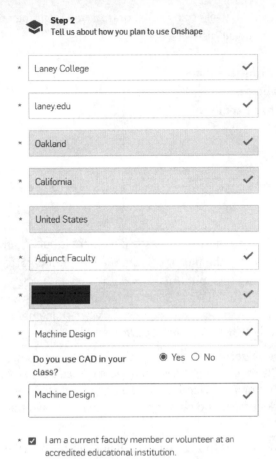

Step 2
Tell us about how you plan to use Onshape

You will need to fill out an additional form about where you go to school or teach.

Then click **CREATE ACCOUNT.**

Be sure to add alerts@onshape.com to your contacts or "whitelist" or their email might end up in your junk folder. If you can't locate the confirmation email in your junk folder or access it, use the Reset Password link to get a new email sent to you after you have added Onshape to your contacts list.

Select the link that is provided in the email.

Set up a password for the account. Be sure to write it down somewhere! Press the Sign Up button and Onshape will immediately launch.

To ensure you are able to communicate properly with Onshape, you can whitelist this domain and IP address:

- **Domain:** outbound-email.onshape.com
- **IP address:** 167.89.77.213

Many corporate environments will whitelist or proxy the network traffic to enforce where users on the network can navigate. In order to use Onshape within this network configuration, the following domains must be added to the "allowed" list: **cad.onshape.com**

In addition, the region of the network as well as any collaborators you might be collaborating with should also be added to the "allowed" list:

- North America: **cad-usw2.onshape.com**
- Europe: **cad-euw1.onshape.com**
- Singapore: **cad-aps1.onshape.com**
- Australia: **cad-aps2.onshape.com**
- Japan: **cad-apn1.onshape.com**

If the network configuration supports domain wildcards, a single entry can be added to the "allowed" list: *.onshape.com

Onshape has two document states: Public and Private. If a document is public, anybody with an Onshape account can view and download the file. If you are the owner or creator of the document, you can share it with other users, but the users must have an Onshape account. Both public and private documents can be shared. Sharing a document allows anyone who has been provided access to make modifications to the document. More than one user can edit a shared document at a time, similarly to how Google Docs works.

If you are not the owner/creator of a document, you cannot share it. If you find a public model that you like, you can send the URL to a friend, so they can see it. Since neither you nor your friend own the model, neither you nor your friend will be able to modify the file. Alternatively, you can make a copy of a public document and own the copy. Once you own the copy, you can share it with anyone you like.

You can transfer ownership of a document to another user. If you are finished working on a project, you can hand it off to another user and let them take over.

Navigating Documents and Workspaces

Estimated Time: 30 minutes

Objectives:

- Getting familiar with the Onshape user interface

1.

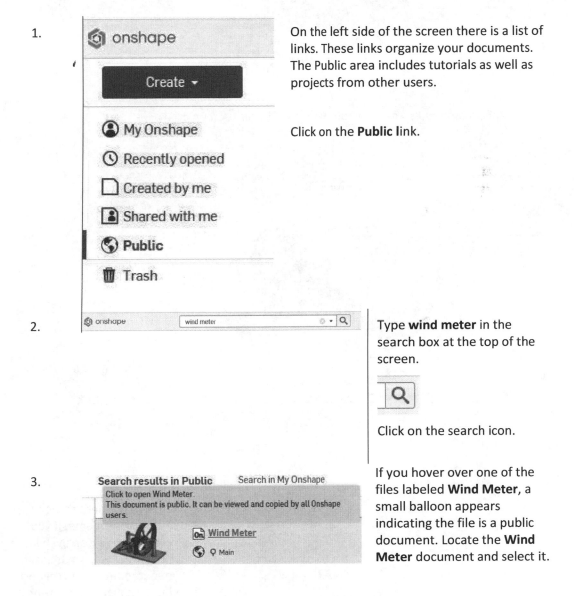

On the left side of the screen there is a list of links. These links organize your documents. The Public area includes tutorials as well as projects from other users.

Click on the **Public l**ink.

2. Type **wind meter** in the search box at the top of the screen.

Click on the search icon.

3. If you hover over one of the files labeled **Wind Meter**, a small balloon appears indicating the file is a public document. Locate the **Wind Meter** document and select it.

4.

When you left click on the link there is a pause while the model is loaded into your workspace.

Take a moment to get familiar with the different screen areas.

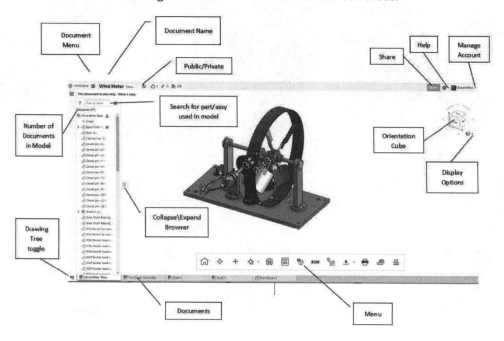

5.

In the upper right corner, you see your user name. If you select the down arrow next to your name, you can manage your account – which allows you to change the password or upgrade to a paid account. You can also view any support tickets.

6.

Action items

Type
☐ Release
☐ Comment

Role
○ Any
◉ Assigned to me
○ Created by me

Status
○ Any
◉ Open
○ Closed

Sort
Oldest first ▼

Action Items allows you to keep track of your work. You may be tasked to review someone else's design or drawing. You may be asked to create a part that is being used in a team project.

If you are a solo user, you can use Action items to help you keep track of where you are in a project.

7.

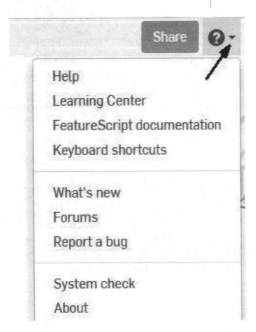

If you select the down arrow next to the Question mark, you are in the Help menu area.

Select **Keyboard shortcuts**.

8.  A nice table of shortcuts appears.

9. Select the Arrow in the top right corner of the keyboard shortcuts window.

A webpage will appear that can be printed out. Simply press Ctrl+P and select the desired printer. You can then pin this up to your cubicle wall or keep it handy for easy reference.

10. Click the **X** icon to close the window.

11. The Share button in the upper right corner will not work when you are in a public document. You will see a little NO symbol if you try to share a public document.

Wind Meter Main

This is a public document.

It is view only. Make a copy

The icon next to the active document name at the top of the screen indicates whether the document is public or private.

12. 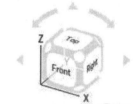 The Orientation Cube located in the upper right of the display window allows you to modify your view easily.

13. 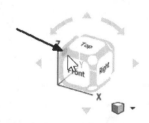 Notice that as you hover over different areas of the orientation cube it will highlight.

Left click to select the shaded area as the new view orientation.

14.

If you left click on the smaller cube (Display Options cube), you will get a shortcut menu where you can change the different display options as well as default orientations.

15.

Instances (77)

📓 Wind Meter Main

⊙ Origin

› Base Plate <.

Belt <1>

Casing ring <1>

In the model browser, you see a list of parts and sub-assemblies. The Base Plate has an icon next to it to indicate it is grounded or fixed in place.

The Base Plate, belt, casing ring, and dowel pins are parts.

16.

Instances (77)

📓 Wind Meter Main

⊙ Origin

› Base Plate <...

Belt <1>

Casing ring <1>

The main assembly also has an icon next to it.

This icon indicates that some of the parts within the assembly are grounded or fixed in place.

17.

Wind Meter Main

⊙ Origin

› Base Plate <...

Belt <1>

Casing ring <1>

Dowel p

Dowel p   Properties...

Dowel p   Hide

Dowel p   Hide other instances

Dowel p   Hide all instances

Dowel p   Isolate...

Dowel p   Make transparent...

Dowel p   Use best available tessellation

Highlight the Dowel pin part.

Right click and select **Isolate**.

18.

All the parts that are not the selected part will become transparent. You may need to orbit the model around to locate the dowel pin.

To rotate the model, use the **ROTATE** tool on the toolbar.

19.

To pan or move the view, select the PAN tool on the toolbar.

The PAN tool doesn't change the position of the model. It adjusts your position relative to the model – like panning a camera.

20.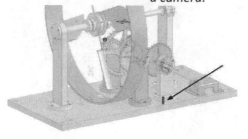

Can you locate the dowel pin?

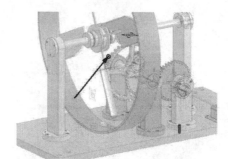

The small dot is the origin.

21.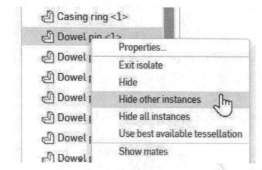

Highlight the Dowel pin.

Select **Hide other instances**.

22.

All the parts are hidden.

You should see the dowel pin, the origin, reference points and reference planes.

23.

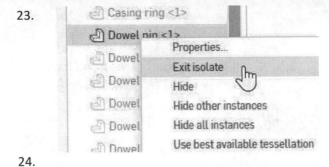

Highlight the Dowel pin.

Right click and select **Exit isolate**.

You still don't see all the other parts.

24.

Right click in the window and select **Show all**.

Visibility of the model is restored.

25.

At the bottom of the screen, there are several tabs. The tabs list the names of the sub-assemblies in the model.
This allows you to quickly navigate to the desired tab.
Left click on **Shaft 1**.

26.

The sub-assembly is displayed.

27.

Instances (5)

📑 Shaft 1

⊙ Origin

⟩ 🔧 Pinion shaft <1>

🔧 Pinion pulley key <1>

🔧 Pulley <1>

🔧 Pinion gear 2 <1>

🔧 Pinion gear key <1>

∨ Mate Features (1)

🔲 Group 1

The model browser lists the parts used in the assembly.

Note that the browser displays (5) instances and there are five parts used in the sub-assembly.

28.

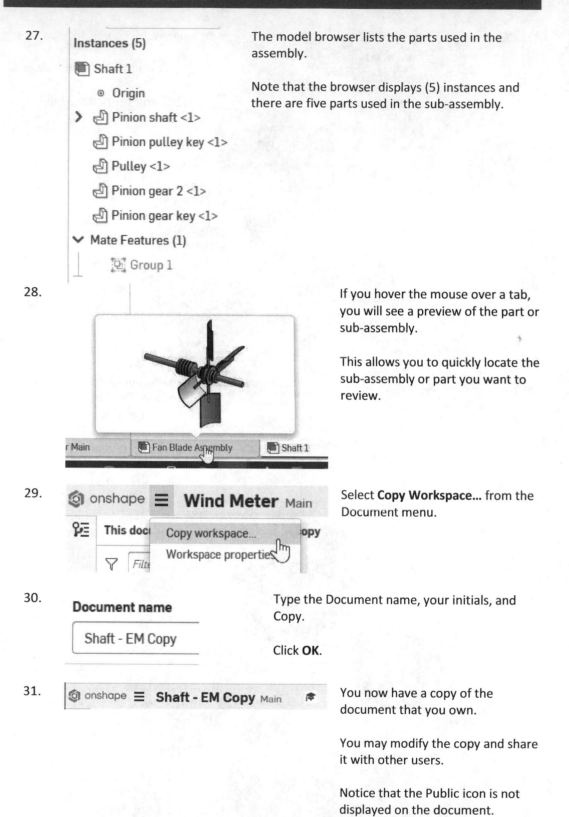

If you hover the mouse over a tab, you will see a preview of the part or sub-assembly.

This allows you to quickly locate the sub-assembly or part you want to review.

29.

Select **Copy Workspace...** from the Document menu.

30.

Document name

Shaft - EM Copy

Type the Document name, your initials, and Copy.

Click **OK**.

31.

You now have a copy of the document that you own.

You may modify the copy and share it with other users.

Notice that the Public icon is not displayed on the document.

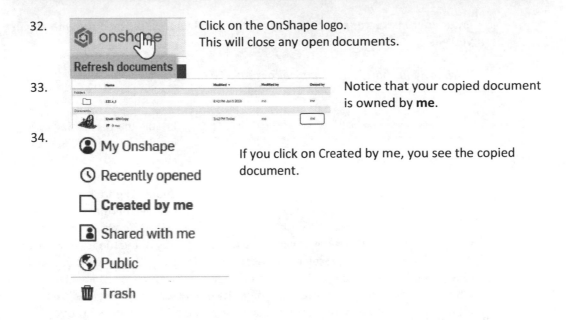

32. Click on the OnShape logo.
This will close any open documents.

33. Notice that your copied document is owned by **me**.

34. Notice that your copied document is owned by **me**.

If you click on Created by me, you see the copied document.

Sharing a Document

Estimated Time: 5 minutes
Objectives:

- Collaboration

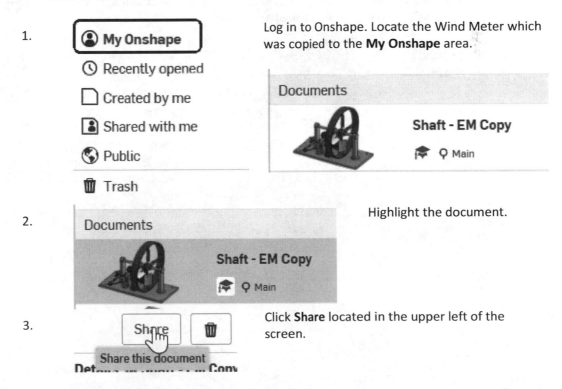

1. Log in to Onshape. Locate the Wind Meter which was copied to the **My Onshape** area.

2. Highlight the document.

3. Click **Share** located in the upper left of the screen.

4. Enter your instructor's email address or the email address of someone with whom you wish to share the document.

Anybody you share a document with is called a Collaborator.

5.  Enable the permissions you would like the receiver to have.
Enable **Copy**.
Enable **Export**.
Enable **Share**.
Enable **Comment**.

If Copy is enabled, the receiver can make their own copy of the model to work on.
If Export is enabled, the receiver can export the file to be 3D printed or used in a different CAD software.
If Share is enabled, the receiver can share the model with anyone else.
If Comment is enabled, the receiver can review and add comments to your document.

6.  In the drop-down list, set the permission to **Can view.**

Notice that when you changed the setting to **Can view**, the recipient can only Comment on the document. All the other permissions are disabled.

If you want to transfer ownership to another user, select **Can Edit** from the drop-down list. Then left click on the **Share** button.

7. Click **Share.**

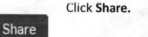

The Collaborator's email is now listed.

You can edit the collaborator's permissions or delete them.

New document shared with you

Hello Elise Moss,

An Onshape document has been shared with you by Susan Moss.

Click to access the document, Shaft - EM Copy

GO TO YOUR ONSHAPE DOCUMENT

This is the email message the collaborator will receive.

| Individuals | Public | Application | Link sharing |

8.

Any Onshape user can view and copy this document.   Make public

☑ Copy ☑ Link document ☑ Export

If you click on the Public tab, this is where you can share your document with the Onshape community.

Do not do this if you are working for a private company or you don't want your work to be copied.

9.

| Individuals | Public | Application | Link sharing |

Clara.io Visualizer ▼ C

Clara.io Visualizer
SolidParts
Onshape BOM for Google Sheets
Maxwell for Onshape

If you click on the Application tab, you see a list of Onshape partners that will import your model.

10.

| Individuals | Public | Application | Link sharing |

Turn on link sharing to allow anyone with the link to view this document. Turn on link sharing

☐ Export

If you click the Link sharing tab, you can create a link that you can email or text to one or more people so they can view and comment on your document.

11.

Close

Click **Close**.

When you first create a document, there are three planes in the display window. The planes are named Front, Top, and Right. These correspond with the front, top, and right views in a 2D drawing.

The planes are displayed as thin rectangles. Planes are infinite, but there really isn't a good way to display an infinite plane. Think of the plane as a limitless piece of paper.

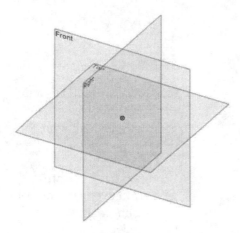

In the browser, under Default geometry, the Origin and the three planes are listed.

∨ Default geometry

    Origin

    Top

    Front

    Right

The origin is the 0,0,0 (x =0, y=0, z=0) point in the 3D universe where you are creating the design.

The origin basically allows you to locate the geometry in relation to 3-dimensional space. If you don't provide a location for the lines, circles, and arcs, it could move around and the results would not be pretty.

Each part is created by drawing a sketch on a plane, then creating a feature from the sketch, and then adding on.

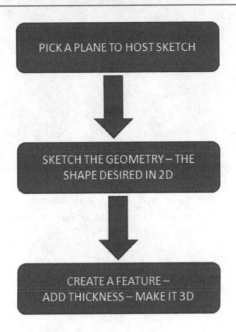

There are four primary feature types:

- Extrude
- Revolve
- Sweep
- Loft

For an Extrude, you place a sketch and then add height or thickness. Common objects that use extrudes are boxes.

For a Revolve, you place a sketch and then revolve it around an axis. An axis can be an edge or line

Common objects that use revolves are wheels and gears.

For a Sweep, you create two sketches. One sketch is the profile and one sketch is the path. The  profile travels along the path to create the 3D shape. The two sketches are usually perpendicular to each other. The path would be on one plane, and the profile would be on a different plane.

Common objects that use sweeps are handles, ducts, and cords.

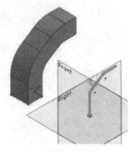

A Loft is also called a blend. A loft is created by placing two or more sketches on offset planes. Each sketch is usually a different shape. The loft blends the sketches together to create a unique 3D feature. Common objects that use lofts are faucets and bottles.

When you model a part, you usually create more than one feature. A part might include an extrude, a revolve, and a loft or it might just be a single sweep. It depends on how complicated the part is. As you work through this text, some parts will be fairly simple and other parts will be quite complex.

When you create a sketch, you draw the geometry – lines, arcs, and circles. Then you use sketch constraints to determine how the elements interact. There are two types of sketch constraints – geometric and dimensional. Geometric constraints are horizontal, vertical, perpendicular, parallel, concentric, coincident, tangent, equal, midpoint, symmetric, etc.

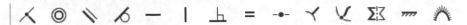

Geometric constraints are located on the sketch ribbon. To add a sketch constraint to geometry, select the element(s) and then select the desired constraint.

Sketches can be under-defined, fully-defined, or over-defined. Under-defined means the sketch does not have all the information – dimensions or geometric constraints – to fully constrain or define all the elements. Fully defined means that all the geometry has been fully designated. Over-defined means that there is duplicate information in the sketch.

When a sketch element is displayed in blue that indicates it is not fully defined. When a sketch element is defined in black, it is fully defined.

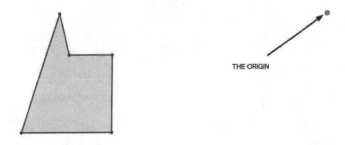

THE ORIGIN

Because the sketch is located in 3-dimensional space, the elements always need to be located in relation to the origin in order to be fully defined. Otherwise, they are just floating in space.

As dimensions and geometric constraints are placed, the geometry changes color.

Black represents geometry that is fully defined, and blue indicates geometry that is under defined.

Some of the elements in this sketch are displayed as blue because it still isn't indicated where it is located in relation to the origin.

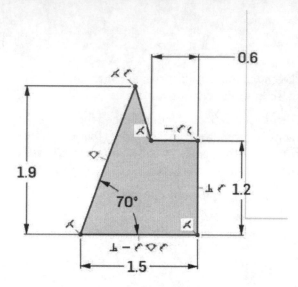

Once the two dimensions are added to indicate where the elements are located relative to the origin, the entire sketch will display in black – fully defined.

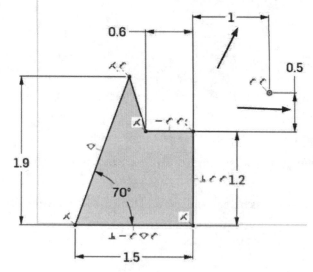

The browser keeps track of the sketches and features as they are placed. You can go back and modify sketches and features at any time. Sketches can be moved to a different plane, if desired as well.

The Onshape document consists of multiple tabs.

Right now there are three types of tabs that can be generated:

- Part Studio – where parts are defined
- Assembly – where parts are assembled
- Drawing – 2D drawings to be annotated

There is no limit to the number of tabs in any single document. So, even with a free account, you could use a single document to store all your parts and not exceed the document limits of a free account.

A Variable Studio is a table. This is useful when creating table-driven parts.

A table-driven part is a part that has similar features, but different sizes.

For example, you can use the variable studio to create screws of different sizes or panels with different sizes and openings.

Notes:

Chapter 2: Signet Ring Project

For the first project, we will design a custom signet ring that can be rapid prototyped. Many schools have their own 3D printers. If you do not have access to a 3D printer at your school or office, you can download your model file in an stl format onto a jump drive and take it to your local office supply store, such as Staples or Office Depot. These stores have 3D printers available and can print your model for a small charge.

This also walks you through how to create the stl file once you are satisfied with your design.

Features covered:
- o EXTRUDE
- o SWEEP
- o FILLET
- o CUT FEATURES
- o IMPORT AN AUTOCAD DWG
- o EXPORT TO STL

Create a New Workspace

Estimated Time: 20 minutes
Objectives:
- Create a New Document
- Create a Sketch
- Use the Viewcube
- Turn off the Visibility of the Reference Planes
- Create an Extrude
- Rename a Feature

1. Log in to Onshape.

2. 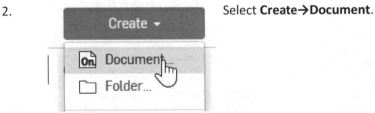 Select **Create→Document**.

3. Name the new document: **Signet Ring**.

   Click **Create**.

   Onshape allows you to assign labels to documents. This makes it easier to organize your documents and search for them. You can also create folders to organize your documents by project or by type. No other user can see the labels you assign to your documents, regardless of whether they are shared or not.

Note that you are in the Part Studio tab. This is the area where you can create and modify parts.

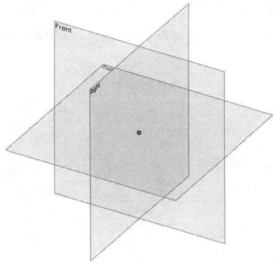

There are three reference planes (Front, Top, and Right) in the display area. There is a small circle where the three planes intersect. This is the Origin.

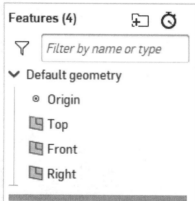

The browser lists the default geometry.

Notice that the reference planes and origin are listed as four features.
You can use the down arrow to collapse the model tree.

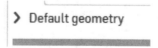

4. Left click on the Top reference plane.

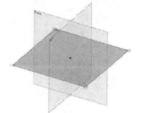

This selects the top reference plane.

You can also select the Top reference plane by left clicking on it in the browser.

5.

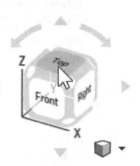

Select the **Sketch** tool on the Part Studio command ribbon.

6.

The browser indicates there is an open sketch. *The Sketch dialog indicates the sketch is being placed on the Top plane.*

Enable **Show constraints** in the Sketch dialog box.

Enable **Show overdefined.**

This will alert you if there are too many constraints added.

7.

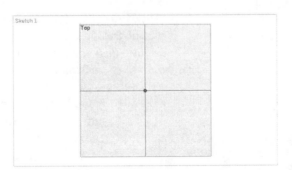

Select the **Top** face of the Orientation cube to reorient the sketch so the view is normal to the sketch.

The view should update.

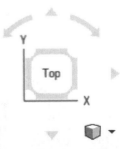

Notice how the Orientation cube indicates the current view orientation

8.

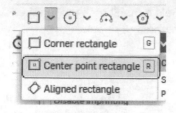

Select the Down arrow next to the rectangle tool.

Select **Center point rectangle**.

9.

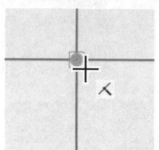

Place your cursor over the origin. Notice how the origin highlights.

A small icon appears to indicate that a coincident constraint will be added to the sketch.

10.

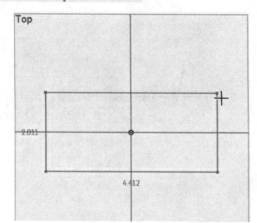

Left click on the origin to start the rectangle. Move the cursor up and to the right to create the rectangle.

Notice that the dimensions change depending on where your cursor is.

Left click to place the top right corner of the rectangle.

11.

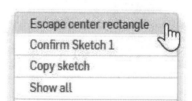

You are still in the Centerpoint rectangle command, so you could place a second rectangle.
Right click and you will see a shortcut menu.
Left click on **Escape centerpoint rectangle**.
You can also Click the ESC key on the keyboard to exit a command.

12.

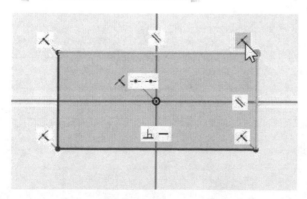

If you hover your mouse over any of the constraint icons, it will highlight the elements belonging to the sketch constraint.

13. 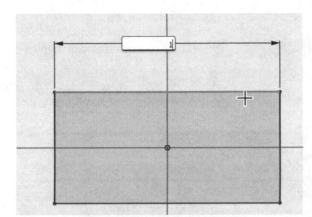 Select the **Dimension** tool on the ribbon.

*The keyboard shortcut for Dimension is **D**.*

14.
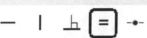

Select the top line.
Left click above the top line to place the dimension.
Change the value to **1 inch**.
Click **ESC** to exit the dimension command.

15. Select the **Equal** constraint from the sketch ribbon.

*The keyboard shortcut for Equal is **E**.*

16.
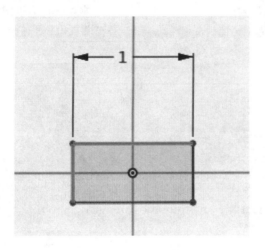

Select the top line of the rectangle.
Select the left vertical line of the rectangle.
Both lines should be highlighted.

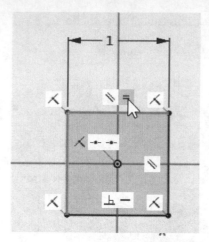

The size of the rectangle will adjust. The color of the rectangle will change to black to indicate it is fully defined.

You will see the equal symbol.
If you hover over the equal symbol the two lines that you selected will highlight.

Click **ESC** on the keyboard.

17.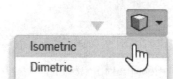

Left click on the Display Options cube.

Select **Isometric**.

18.

In the browser:
Right click on the **Top** plane.
Select **Hide all planes**.

The visibility of the planes is turned off.
All that is visible is the origin, the Sketch plane and the center rectangle.

19.

Select the **Extrude** tool from the command ribbon.

20.

Set the Depth to **.125 in**.
Enable **Draft**.
Set the Draft to **3 deg**.
Left click on the arrow next to the draft text field.

This changes the direction of the draft. Observe how the model updates as you modify the inputs.

Left click on the **Green Check** in the feature dialog.

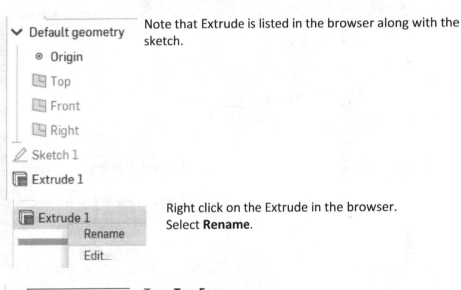

Note that Extrude is listed in the browser along with the sketch.

21.

Right click on the Extrude in the browser.
Select **Rename**.

22.

Type **Top Face**.
Click ENTER.

The Extrude is now renamed.

It's a good idea to name your features to make it easier to locate in your model when you need to make modifications.

23.  Left click on the Onshape logo in the upper left of the screen to close the document.

Assign a Label

Estimated Time: 5 minutes

Objectives:

- Create Labels
- Search Labels

1.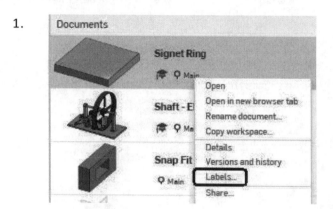

   Right click on the **Signet Ring** document.

   Select **Labels**.

2. 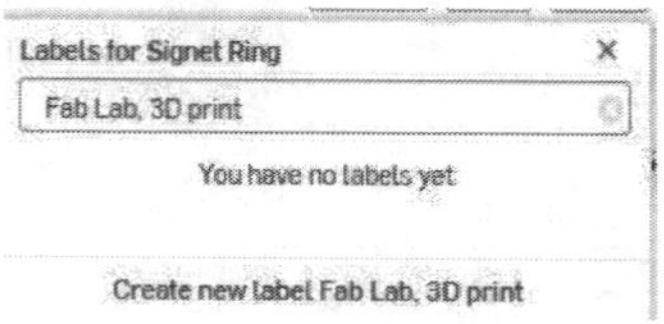 Type **Fab Lab, 3D Print**.

   Click **Create new label** at the bottom of the dialog box.

3. 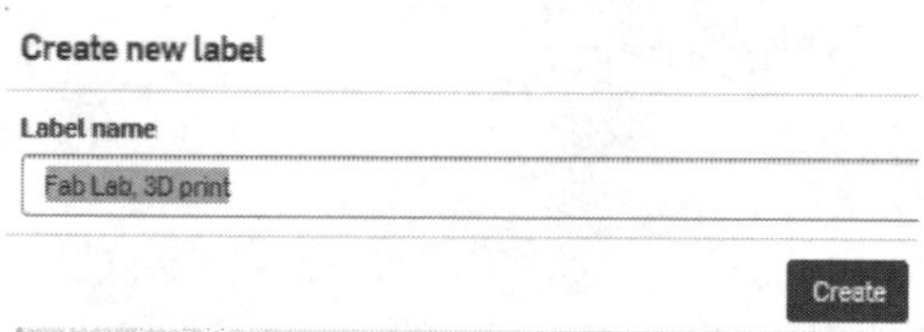 Click **Create**.

   The labels now appear next to your document.

2-8

4.

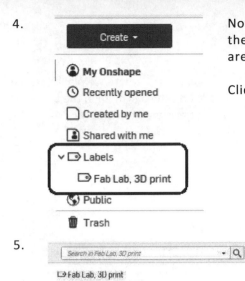

Notice on the left side of the display window there is a category called Labels. The new labels are listed.

Click on the **Fab Lab, 3D print** label.

5.

The document with the selected label is listed.

Create a Sweep

Estimated Time: 45 minutes
Objectives:
- Modify a Document
- Create a Sketch
- Create a Sweep

1.

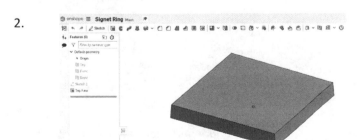

Sign into the Onshape account.

Left click on the **Signet Ring** document.

2.

To zoom out, use the scroll wheel on the mouse.
To orbit or rotate the model, Click down on the right mouse button.
To pan (move from side to side or up and down without changing the camera distance to the model), click down the Ctrl key and the right mouse button at the same time.
To Zoom Extents, simply

double click the mouse wheel.

3.

To make the reference planes visible, click on the > symbol in the browser next to Default geometry.

4.

Highlight the Front plane in the browser.
Right click and select **New sketch.**

5.

Enable **Show Constraints**.

6.

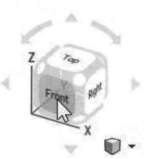

Select the front plane on the Orientation cube.

When the view changed, you may have exited the sketch.

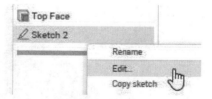

If that happened, highlight the Sketch in the browser. Right click and select **Edit**.

7.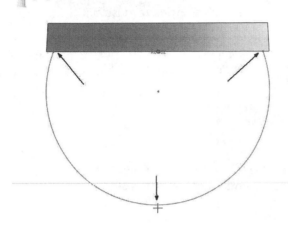

Select the **3 point arc** tool from the Sketch tools ribbon.

8.

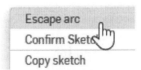

Draw the arc as shown.
Place the left end point on the left side bottom of the extrude.
Place the right end point on the right side bottom of the extrude.

Move the mouse down so the center of the arc is below the extrude.

Left click to complete the arc.

9.

Right click and select **Escape arc**.

This exits the command.

Escape arc
Confirm Sketc
Copy sketch

10.

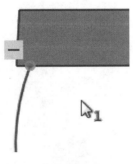

Select the left end point.

A 1 appears next to the cursor to indicate one selection has been made.

11.

Select the bottom edge of the extrude.

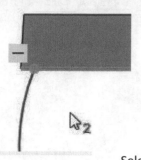

A 2 appears next to the cursor to indicate two selections have been made.

12.

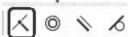

Select the **Coincident** constraint.

Check to see if a coincident symbol appears to indicate the constraint has been placed.

If you don't see any constraint icons, verify that Show constraints is enabled in the Sketch box.

13.

Repeat on the right side.
Select the right end point.
Select the bottom of the extrude.
Then select the **Coincident** tool.

14.

Select the **Dimension** tool from the ribbon.

15.

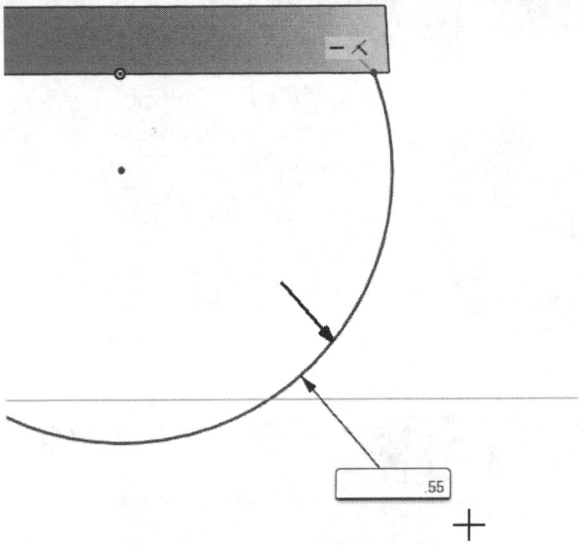

Select the arc.
Set the dimension to **0.55**.

.55

16.

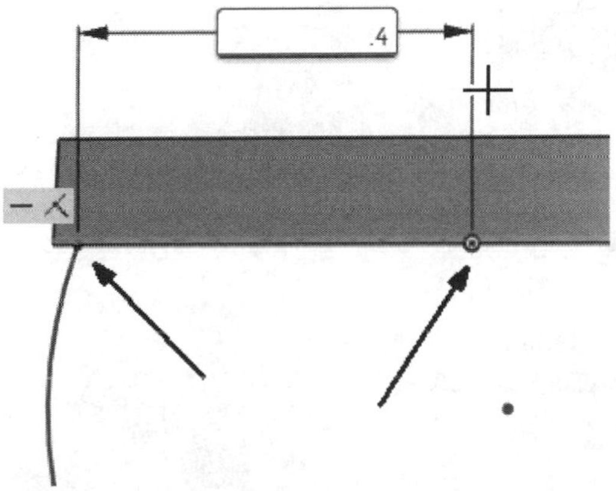

Place a 0.4 dimension
between the left end point
and the origin.

.4

17.

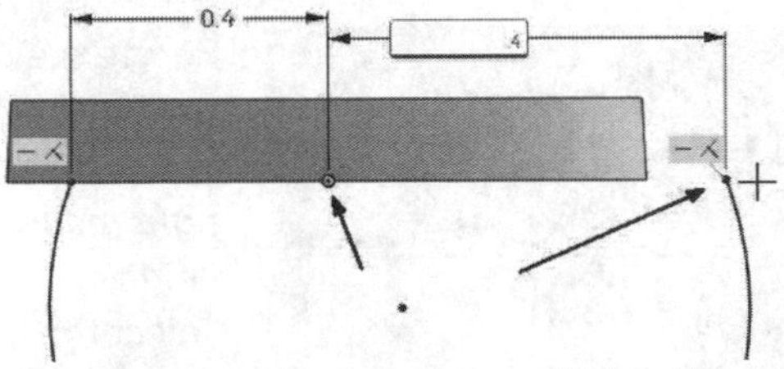

Place a 0.4 dimension
between the right end point
and the origin.

0.4   .4

18.

Right click and select **Escape dimension**.

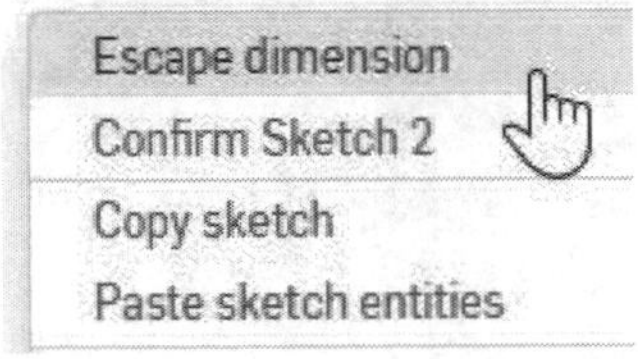

Escape dimension
Confirm Sketch 2
Copy sketch
Paste sketch entities

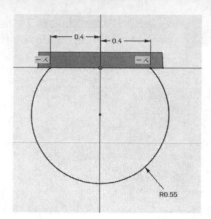

The color of the sketch should be black, indicating that it is fully defined.

If your sketch isn't black, check to see if you are missing the coincident constraints or a dimension.

19.

Select the **Green check** symbol.

This exits the sketch.

20.

Highlight the sketch in the browser.
Right click and select **Rename**.

21.

Rename the sketch **Path**.
Click **ENTER**.

22.

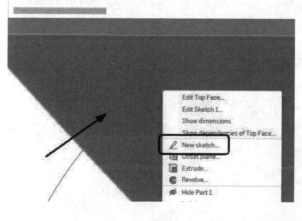

Rotate the part so you can see the bottom of the extrude.
To rotate the view, Click the right mouse button down and hold.
Select the bottom face.
Right click and select **New sketch**.

23.

Enable **Show constraints**.

24.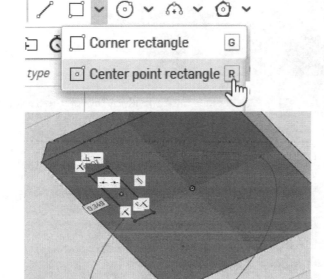

Select the **Center point rectangle** tool.

The shortcut key for the centerpoint rectangle is R.

25.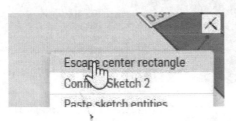

Place the rectangle next to the left end point of the arc.

26.

Right click and select **Escape center point rectangle**.

This exits the command.

27.

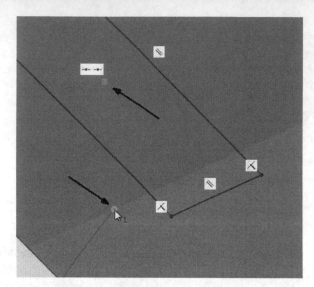

Left click to select the center point of the rectangle.
Left click to select the left end point of the arc.

28.

Select the **Coincident** tool on the Sketch ribbon. The rectangle should shift over.

29.

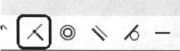

Select the **Dimension** tool on the Sketch ribbon.

30.

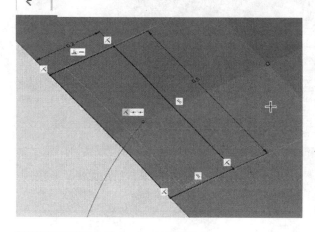

Place a **0.2** dimension for the width of the rectangle.
Place a **0.5** dimension for the length of the rectangle.

The sketch should change color to black to indicate it is fully defined.

31.

Sketch 2   ✓ ✕

Sketch plane
Face of Sketch 1   ✕

☑ Show constraints
☑ Show overdefined

Click the Green Check to exit the sketch.

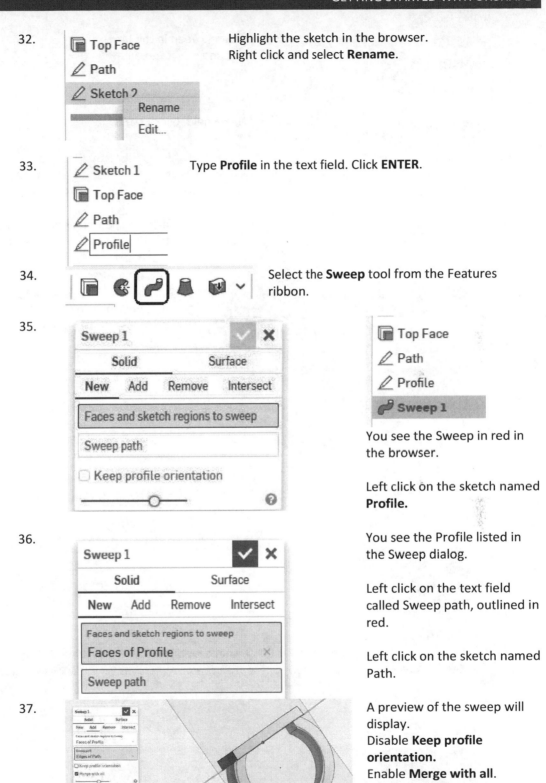

32. 📖 Top Face
    ✎ Path
    ✎ Sketch 2
           Rename
           Edit...

Highlight the sketch in the browser.
Right click and select **Rename**.

33. ✎ Sketch 1
    📖 Top Face
    ✎ Path
    ✎ Profile

Type **Profile** in the text field. Click **ENTER**.

34. Select the **Sweep** tool from the Features ribbon.

35. Sweep 1 ✓ ✗

    Solid   Surface

    New   Add   Remove   Intersect

    Faces and sketch regions to sweep

    Sweep path

    ☐ Keep profile orientation

    📖 Top Face
    ✎ Path
    ✎ Profile
    🦯 Sweep 1

    You see the Sweep in red in the browser.

    Left click on the sketch named **Profile.**

36. Sweep 1 ✓ ✗

    Solid   Surface

    New   Add   Remove   Intersect

    Faces and sketch regions to sweep
    Faces of Profile   ✗

    Sweep path

    You see the Profile listed in the Sweep dialog.

    Left click on the text field called Sweep path, outlined in red.

    Left click on the sketch named Path.

37. A preview of the sweep will display.
    Disable **Keep profile orientation.**
    Enable **Merge with all**.
    Left click on the Green Check to complete the sweep.

38. Highlight the Sweep in the browser.
Right click and select **Rename**.

39. Type **Shank** in the Name field. Click **ENTER**.

40.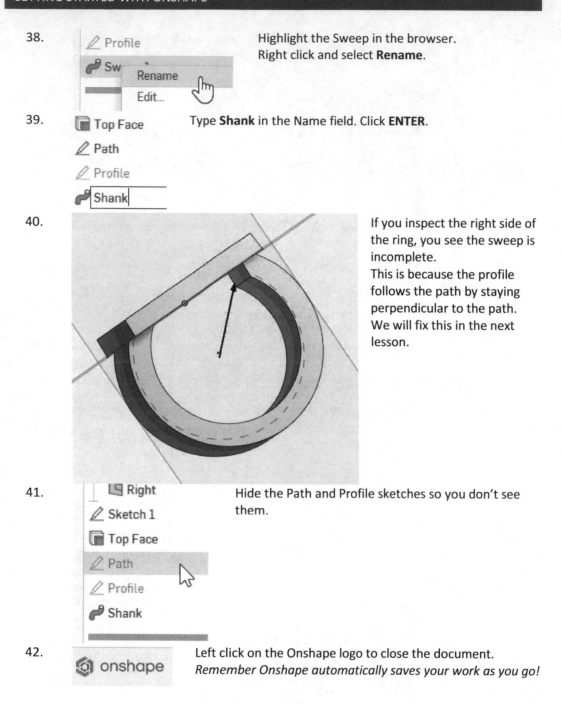
If you inspect the right side of the ring, you see the sweep is incomplete.
This is because the profile follows the path by staying perpendicular to the path.
We will fix this in the next lesson.

41. Hide the Path and Profile sketches so you don't see them.

42. Left click on the Onshape logo to close the document.
Remember Onshape automatically saves your work as you go!

Project Edges in a Sketch

Estimated Time: 25 minutes

Objectives:

- Modify a Document
- Create a Sketch
- Project/Convert an Edge
- Extrude to a Face in both directions

1.

   Sign into the Onshape account.
   Left click on the Signet Ring document.

   Did you notice that the preview image updates as you add more features?

2. Highlight the **Front** plane in the display window.
   Right click and select **New sketch**.

   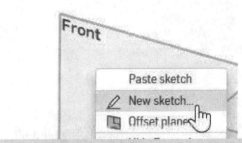

3.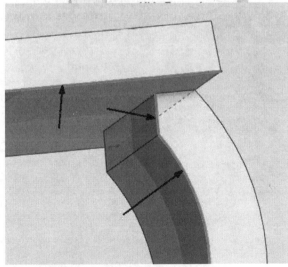

   Left click on the bottom edge of the extrude, the small vertical edge indicated, and the inside arc.

   Hold down the right mouse button to orbit the model so you can make the selections.

4.

   Select the **Use (ProjectConvert)** icon on the Sketch ribbon.
   This copies the selected geometry into the active sketch.

5.

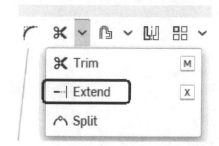

Locate and select the **Extend** tool underneath the Trim icon on the Sketch ribbon.

The shortcut key for Extend is X.

6.

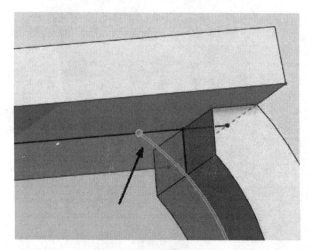

Select the end point of the arc and drag the end point above the horizontal line.

Left click to indicate the end of the extension.

7.

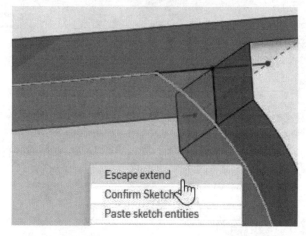

Note that the arc's endpoint is extended to the horizontal line.
Right click and select **Escape extend**.

8.

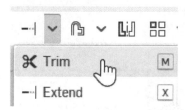

Select the **Trim** tool.

The shortcut key for Trim is M.

9.

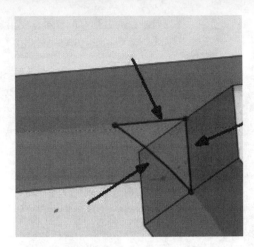

Select the lines and the outside arc, so that all that remains are the two lines and small arc segment.

When using Trim, click what you want to be deleted.

10.

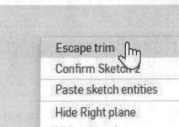

Right click and select **Escape trim**.

11.

There is a small line segment on the left.

Click to select it.
Then use the **DELETE** key on the keyboard to delete.

12.

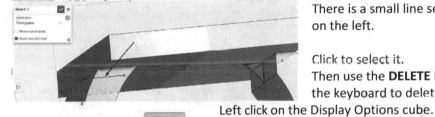

Left click on the Display Options cube.

Select **Isometric.**

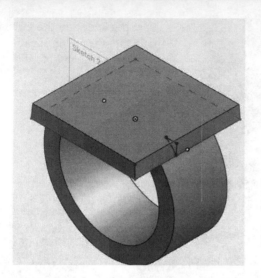

The only items in the sketch should be the two lines and small arc.

If you see any remaining geometry, window around it and select Delete on the keyboard.

13.   Select the **Extrude** tool from the ribbon.

14.

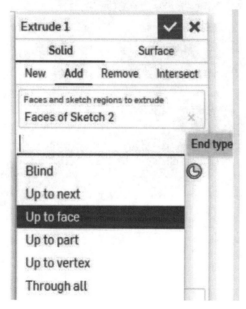

Select **Up to face** for the first direction.

15.

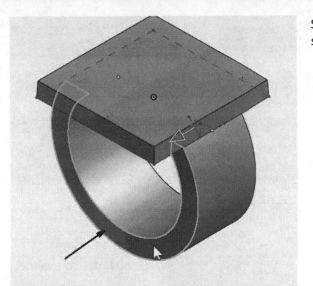

Select the front face of the sweep/shank.

16.

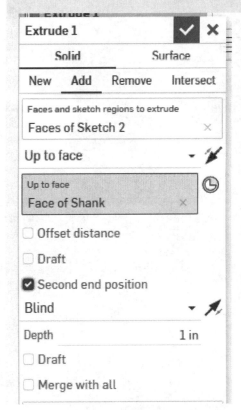

Note the face of sweep is indicated as the termination for the first direction.

Enable **Second end position**.

17.

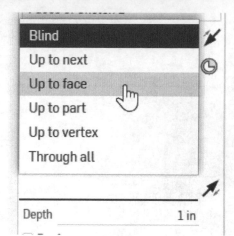

Select **Up to face** for the second end position.

18.

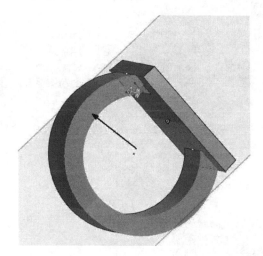

Rotate the model using the right mouse button.
Select the back face of the sweep/shank.

19.

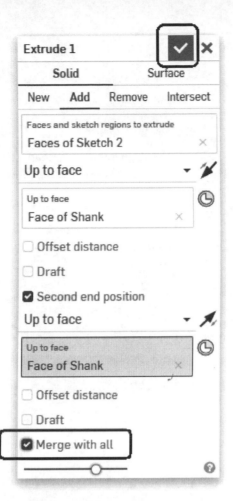

Enable **Merge with all**.
Inspect the preview to see how the extrude is going to look.
Click the **Green check** to complete the extrude.

20.

Inspect the ring to see if the gap has been filled in correctly.

21.

Left click on the Onshape logo to close the document.

Add Fillets

Estimated Time: 10 minutes
Objectives:
- Adding Fillets
- Changing the display settings

1.
Sign into the Onshape account.
Left click on the Signet Ring document.

2.
Left click on the Display Options cube.
Select **Hidden edges visible** to change the model appearance.

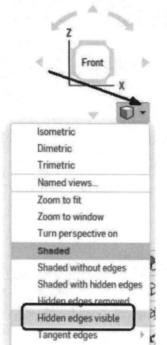

The display will update.

3.
Select the **Fillet** tool from the Features ribbon.

Fillets are a CHILD feature – they must be added to a PARENT feature. You need an extrude/sweep/revolve before you can place a fillet.

The shortcut key for FILLET is SHIFT+F.

4. 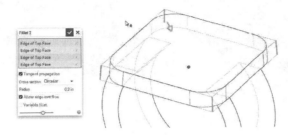 Select the four corner edges
of the top face.
Set the value of the radius to
0.2 in.
Enable **Tangent propagation.**
*Note that you do not need to
rotate the model with the
hidden edges showing.*
Select the Green check to
accept the fillets assigned.

*Notice that a number appears next to the cursor to indicate how many edges have
been selected.*

5. Select the **Fillet** tool from the Features ribbon.

6. 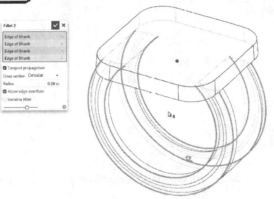 Set the value of the fillet to
0.06 in.
Select the outside and inside
edges of the ring's shank.
Select the Green check to
accept the fillets assigned.

*Notice that four selections
have been made.
Verify that you selected Edge
of Shank and not Face of
Shank or you may see an
error.*

7. Select the **Fillet** tool from the Features ribbon.

8.  Set the value of the fillet to
0.03 in.
Select the top face.
Select the Green check to
accept the fillets assigned.

9.

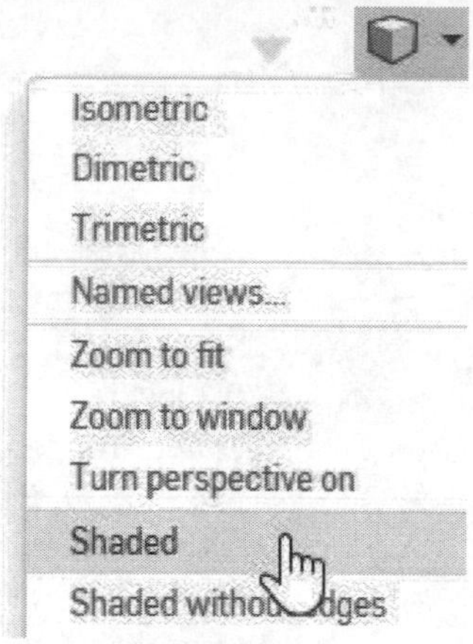

Isometric
Dimetric
Trimetric
Named views...
Zoom to fit
Zoom to window
Turn perspective on
Shaded
Shaded without edges

Left click on the Display Options cube and set the display to **Shaded.**

10.

Inspect your signet ring.

11.

onshape

Left click on the Onshape logo to close the document.

Assign Material and Change Appearance

Estimated Time: 5 minutes

Objectives:

- Change part appearance
- Assign Material

1.

Sign into the Onshape account.
Left click on the Signet Ring document.

2.

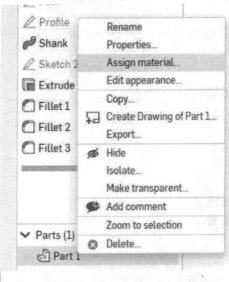

In the lower panel of the browser, locate where it says **Part 1**.
Right click and select **Assign material**.

3.

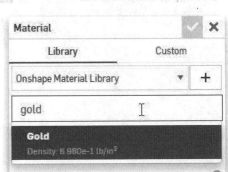

Type gold In the search field.

Click on the material: **Gold** to select.
Then click the Green Check to accept.
Note that this doesn't change the appearance of the ring.

4.

In the browser, locate where it says Part 1.

Right click and select **Edit Appearance.**

5.

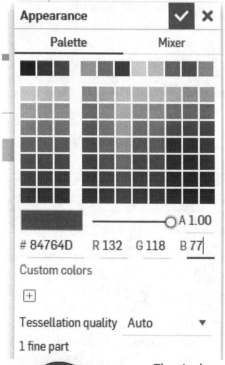

Type **132** in the R field.
Type **118** in the G field.
Type **77** in the B field.
These are the Red, Green, and Blue color values corresponding to Pantone's metallic Gold.
Green Check to accept.

The ring's appearance updates.

6.

Left click on the Onshape logo to close the document.

Branching

Estimated Time: 5 minutes

Objectives:

- Create different versions of the part

1.

   Sign into the Onshape account.
   Left click on the Signet Ring document.

2.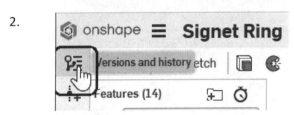

   Select the **Version** tool from the menu.

3.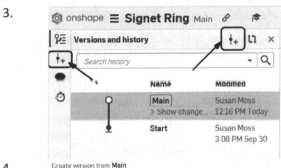

   Select **Create version**.

4.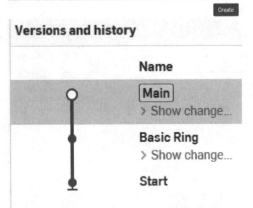

   In the Name field, type **Basic Ring.** In the Description field, type **Ring with no decoration.**

   Click **Create**.

   You now see the Basic Ring listed as a version.

5.

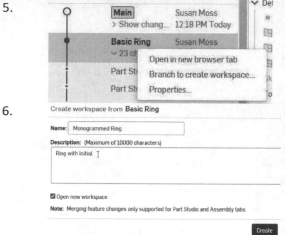

Highlight the **Basic Ring** version.

Right click and select **Branch to create workspace**.

6.

Click **Create**.

In the Name field, type **Monogrammed Ring.**
In the Description field, type **Ring with Initial.**
Click **Create**.

A branch appears from the Basic Ring to the Monogrammed Ring.

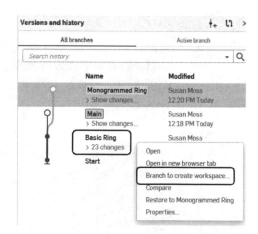

7.

Highlight the **Basic Ring** version.
Right click and select **Branch to create workspace**.

8.

Create workspace from **Basic Ring**

Name: Etched Ring

Description: (Maximum of 10000 characters)

Ring with Etching

☑ Open new workspace

Note: Merging feature changes only supported for Part Studio and Assembly tabs

Create

In the Name field, type **Etched Ring.**
In the Description field, type **Ring with Etching.**
Click **Create**.

Versions and history

All branches

Search history

Name

Etched Ring
> Show changes...

Monogrammed Ring
> Show changes...

Main
> Show changes...

Basic Ring
> 23 changes

Start

You now see two branches coming from the Basic Ring version.

9.

Close the document.

Add Embossed Text

Estimated Time: 20 minutes

Objectives:

- Add Text
- Extrude – Remove (Create Cut)

1.

Sign into the Onshape account.
Left click on the Signet Ring document.

Notice the preview now displays the active branch.

2.

Click on the **Version** icon, if the panel is closed.

Left click on the **Monogrammed Ring** version to activate it.

3.

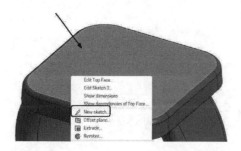

Verify the version name appears above the ribbon next to the Document name.

4.

Left click on the top of the ring. Right click and select **New sketch.**

5.

Enable **Show constraints**.

| Sketch 3 | ✓ | ✕ |

Sketch plane
Face of Top Face ✕

☑ Show constraints
☑ Show overdefined

6.

Orient the model to the **Top** view using the Orientation cube.

7.

Select the **Text** tool from the Sketch ribbon.

8.

Draw a rectangle on the top face to indicate where the text will be placed.

9.

Text

Roboto Slab ▼

Select the desired font from the drop-down list.

*I selected **RobotoSlab**.*
If you are going to 3D print this part, the plainer the font the better. Fancy texts may not provide the desired result.

10.

Type your initial.
Enable **Bold.**
Click the Green Check to finish the sketch.

11.

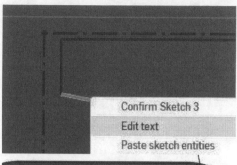

To modify the text, select it.
Right click and select **Edit text**.

12.

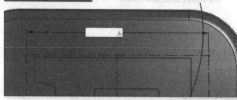

Use the Dimension tool to add a
dimension of **0.5 in** to the top
construction line of the text control
box.

13.

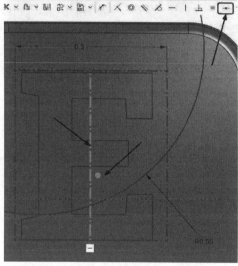

Select the vertical construction line that
goes through the text.
Left click on the Origin in the browser.
Left click on the **midpoint** constraint on
the sketch ribbon.

14.

The text will change to black to indicate that it is fully defined, and it will shift so that it is centered on the top face.

Select the **Green Check** to exit the sketch.

15. Select the **Extrude** tool.

16.

Extrude 2	✓	✗

Solid   Surface

New   Add   **Remove**   Intersect

Faces and sketch regions to extrude
Faces of Sketch 3   ✗

Blind

Depth   0.06 in

☐ Draft

☐ Second end position

☑ Merge with all

Left click on the word **Remove** to cut into the top face.
Select the sketch with the initial from the browser.
Set the distance to **0.06 in**.
Enable **Merge with all**.
Select the **Green Check.**

17. Extrude 2
    Rename

Select the Extruded Cut in the browser. Right click and select **Rename.**

18. ✎ Sketch 3

    Initial

Type **Initial** and Click ENTER.

19.

Inspect the signet ring.

20.   Left click on the Onshape logo to close the document.

Insert an AutoCAD Drawing into a Sketch

Estimated Time: 30 minutes

Objectives:

- Import an AutoCAD file into your account
- Add Text
- Remove Extrude

1.

Sign into the Onshape account.
Left click on the Signet Ring document.

Notice the preview displays the active branch.

2.

Versions and history

All branches

Name

○ **Monogrammed Ring**
  > Show changes...

● **Etched Ring**
  > Show changes...

○ **Main**
  > Show changes...

● **Basic Ring**
  > Show changes...

● **Start**

Click on the **Version** tool on the menu.
Left click on the **Etched Ring** version to activate it.

3.

Signet Ring Etched Ri...

Etched Ring

history

Verify that the Etched Ring version is active.

4.

Inspect the ring to verify that it is the correct version.

5.

Features (14)

∇ *Filter by name or type*

∨ Default geometry

  ◉ Origin

  🔲 Top

  🔲 Front

  🔲 Right

✏ Sketch 1

🔲 Top Face

✏ Path

✏ Profile

🦴 Shank

✏ Sketch 2

🔲 Extrude 1

◻ Fillet 1

◻ Fillet 2

◻ Fillet 3

Notice that the initial feature created in the Monogrammed version of the ring is not listed in the browser.

6.

➕ 🔲 Part Studio 1   📄 Assembly 1

Select the + tab in the lower left of the screen.

7.

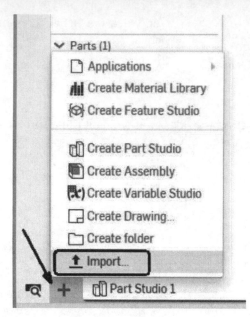

Select **Import**.

8.

Locate the *signet ring.dwg* file.
This file can be downloaded from the publisher's website at sdcpublications.com.
Click **Open**.

A message will appear indicating it was imported correctly.

A tab will be added to indicate that the drawing has been loaded into the active document.

9.

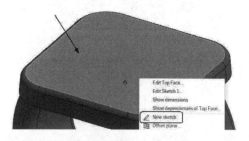

Select the top face for a new sketch.

10.

Select **Insert DXF or DWG** from the Sketch ribbon.

11.

Units Inch ▼

☐ Use file origin position

Set Units to **Inch** in the bottom of the dialog.

12.

Insert a DXF or DWG file

Current document   Other

Signet Ring
↳ Etched Ring

Search DXF or DWG files

signet ring.dwg

Left click on *the signet ring.dwg* name.

The AutoCAD drawing will automatically be placed in the sketch. The origin of the AutoCAD drawing is automatically aligned with the origin of the sketch.

13. Select the **Extrude** tool.

14.

Extrude 2 ✓ ✕
Solid   Surface
New   Add   Remove   Intersect
Faces and sketch regions to extrude
Faces of Sketch 4
Blind ▼ ↗
Depth   0.06 in
☐ Draft
☐ Second end position
☑ Merge with all

Select the **Remove** option.
Set the Termination to **Blind.**
Set the Depth to **0.06 in**.
Enable **Merge with all.**
The face outline is automatically selected.

Click the **Green Check**.

The face outline is placed.

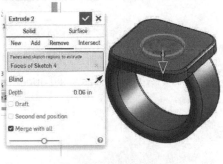

15.

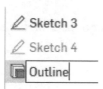

Highlight the Extruded cut.
Right click and select **Rename.**

16.

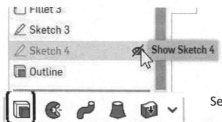

Type **Outline.**
Click ENTER.

17.

Click on the **Show** icon to display the imported sketch.

18.

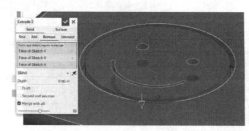

Select the **Extrude** tool from the ribbon.

19.

Select the **Remove** option.
Set the Termination to **Blind.**
Set the Depth to **0.06 in**.
Enable.
Merge with all.
Select the two eyes and the mouth.

Be careful not to select the top face!
If you do, delete in the list box.

Click the **Green Check**.

The etching should appear as shown.

20.

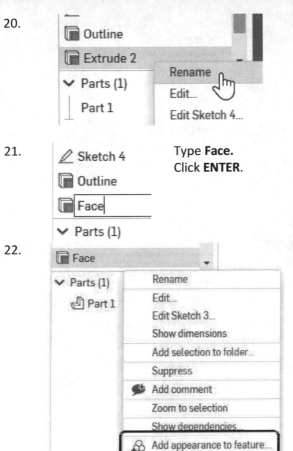

Highlight the Extruded cut.
Right click and select **Rename.**

21.

Type **Face.**
Click **ENTER.**

22.

Highlight the Face feature.
Right click and select **Add appearance to feature**.

23.

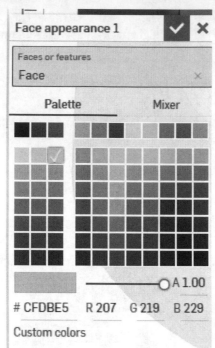

Pick a color to apply to the eyes and mouth.

Click **Green check**.

24.

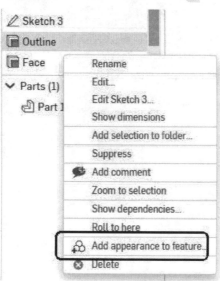

Highlight the Outline feature.
Right click and select **Add appearance to feature**.

25.

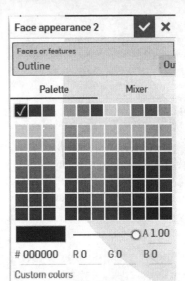

Pick a color to apply to the outline.

Click **Green check**.

26.

Inspect the completed ring.

27.

Close the document.

Compare Versions

Estimated Time: 5 minutes

Objectives:

- See differences between branches of a model

1.

Sign into the Onshape account.
Left click on the Signet Ring document.

Notice the preview displays the active branch.

2.

Click on the **Version** tool on the menu.

3.

Click **Compare**.

4.

Highlight the Etched Ring and the Monogrammed versions.

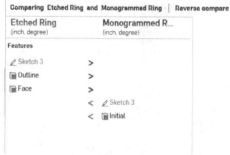

5.

Compare shows which sketches and features are different in each model.

6.

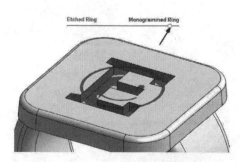

In the display window drag the small ball on the slider from the left to right to see the ring go from the etched version to the monogrammed version.

7.   Close the document.

Extra: *Create a third version of the ring using the decoration of your choice.*

Export to Stl

Estimated Time: 5 minutes

Objectives:

- Create an stl file of an Onshape document

STL files are used to create 3d prints. Once you have the STL file you can send it to a 3D printer, or you can take the file to most office supply stores to have them create the 3D print.

1.      Sign into the Onshape account.
                                     Left click on the Signet Ring document.

                                     Notice the preview displays the active branch.

2.      Click on the **Version** tool on the menu.
                                     Left click on the Monogrammed ring to make it the active document.

3.   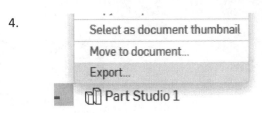   Close the Versions dialog by clicking on the small x in the right corner.

4.   Select as document thumbnail
       Move to document...   Right click on the Part Studio tab and select **Export**.
       Export...
       Part Studio 1

5.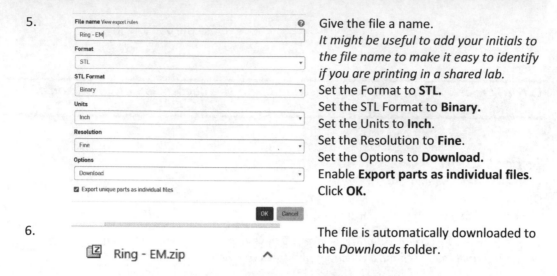

Give the file a name.

It might be useful to add your initials to the file name to make it easy to identify if you are printing in a shared lab.

Set the Format to **STL.**

Set the STL Format to **Binary.**

Set the Units to **Inch**.

Set the Resolution to **Fine**.

Set the Options to **Download.**

Enable **Export parts as individual files**.

Click **OK.**

6.

Ring - EM.zip

The file is automatically downloaded to the *Downloads* folder.

7. You can then locate the file and email it to your instructor or lab to be printed.

8. Close the document.

Chapter 3: Angled Plate Project

Estimated Time: 45 minutes

Objectives:

- Revolve
- Extrude to Face
- Fillet
- Edit Appearance
- Circular Pattern

This part will be used to create a drawing with an auxiliary view.

1.

Sign into the Onshape account.

Select the **Create →Document.**

2.

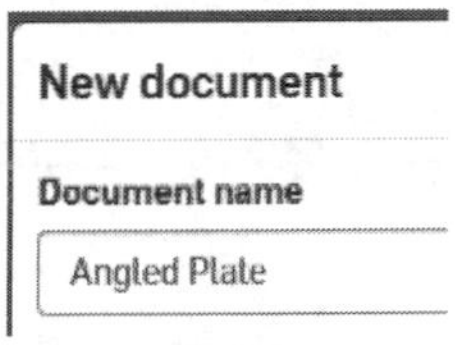

Name the document **Angled Plate**.

Click **Create**.

3.

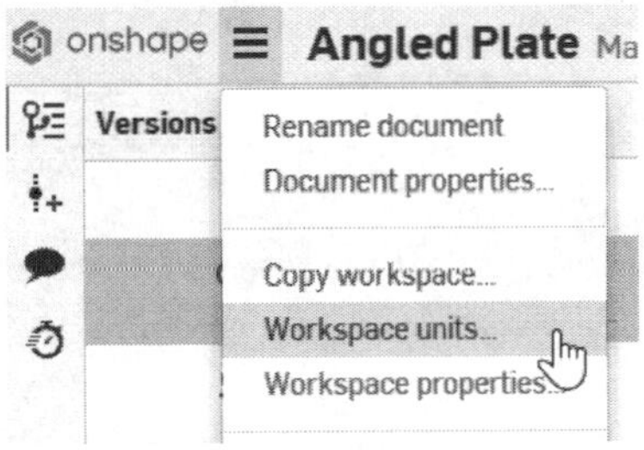

Click on the Document menu.

Click **Workspace units**.

4.

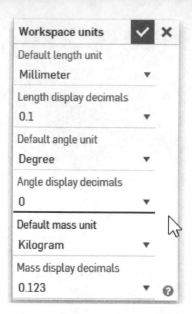

Set the Length to **Millimeter.**
Set the Length display decimals to **0.1.**
Set the Angle display decimals to **0.**
Set the Mass to **Kilogram.**

Click the Green check.

5.

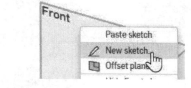

Select the **Front** plane for a **New sketch.**

6.

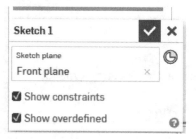

Enable **Show constraints**.

7.

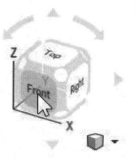

Set the Orientation cube to **Front.**

8.

Select the **Line** tool from the ribbon.

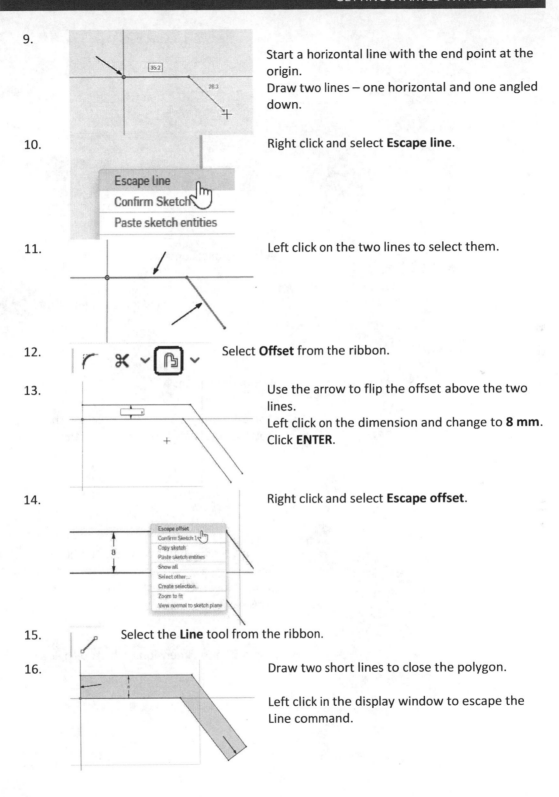

9. Start a horizontal line with the end point at the origin.

Draw two lines – one horizontal and one angled down.

10. Right click and select **Escape line**.

11. Left click on the two lines to select them.

12. Select **Offset** from the ribbon.

13. Use the arrow to flip the offset above the two lines.

Left click on the dimension and change to **8 mm**. Click **ENTER**.

14. Right click and select **Escape offset**.

15. Select the **Line** tool from the ribbon.

16. Draw two short lines to close the polygon.

Left click in the display window to escape the Line command.

17.

Select the two angled lines.

18.

Add a parallel constraint.

Notice that perpendicular constraints have already been added.

19.

Select the **Dimension** tool from the ribbon.

20.

Add a 130° angle dimension between the bottom horizontal line and the bottom angled line.

21.

Add a 50 mm dimension to the top horizontal line.
If the left vertical line shifts, add a vertical constraint.

22.

Add a 50 mm dimension to the top angled line.

23.

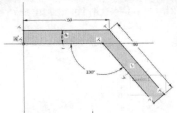

If the short end line on the angled section shifts, add a perpendicular constraint.

The sketch should be fully defined.

24.

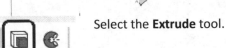

Select the **Extrude** tool.

25.

Select **New.**
Enable **Symmetric.**
Set the Width to **30 mm.**
Click the Green check.

26.

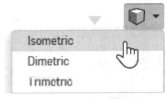

Use the Display Options cube to switch to an **Isometric** view.

27.

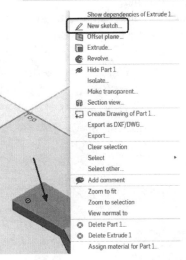

Select the top face for a new sketch.

28.

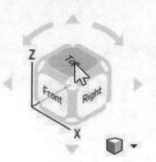

Select the top face on the orientation cube.

29.

Select the **Line** tool.

30.

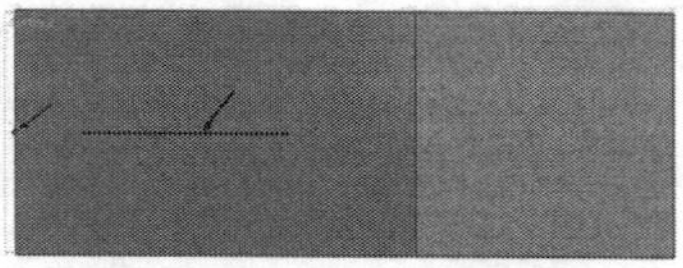

Place a horizontal line in line with the origin.

Right click and **Escape line**.

31.

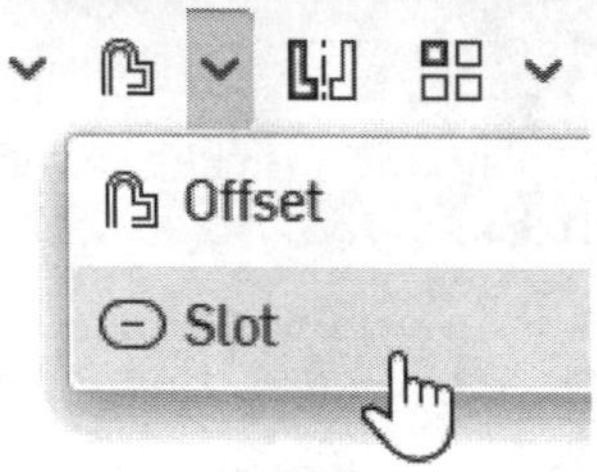

Select the **Slot** tool below Offset on the ribbon.

Offset

Slot

32.

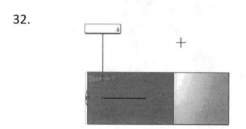

Select the line.
Modify the dimension to **6 mm.**
Right click the mouse twice to place the slot.

33.

Select the **Dimension** tool from the ribbon.

34.

Add a dimension to locate the start center point of the slot **15 mm** from the left edge.
Add a **20 mm** dimension for the length of the slot.

The sketch should display as fully defined.
If it doesn't check that you have a horizontal constraint between the origin and the center line.

35.

Switch to an **Isometric** view using the Display Options cube.

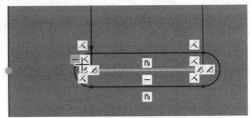

Isometric

Dimetric

Trimetric

36.

Select the **Extrude** tool.

37.

Select **Remove.**
Set the Termination to **Through all.**
Enable **Merge with all**.
Click the Green check.

Extrude 2

Solid   Surface

New   Add   Remove   Intersect

Faces and sketch regions to extrude
Faces of Sketch 2

Through all

☐ Symmetric

☐ Draft

☐ Second end position

☑ Merge with all

38.

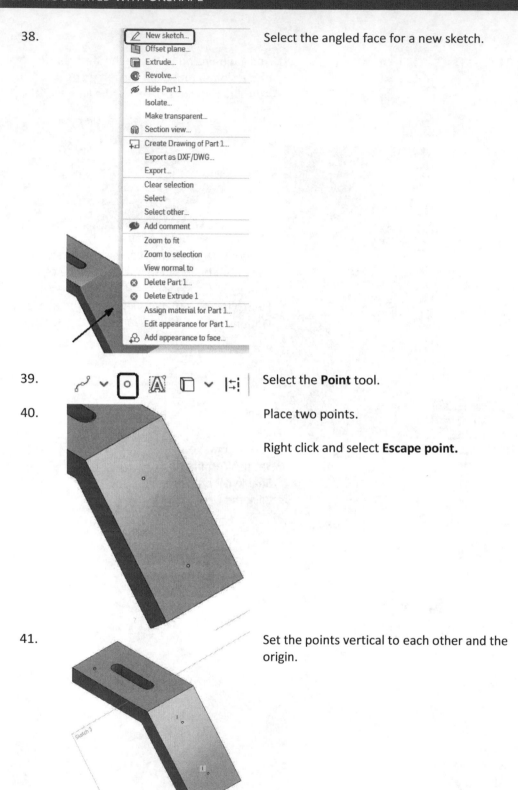

Select the angled face for a new sketch.

39.

Select the **Point** tool.

40.

Place two points.

Right click and select **Escape point.**

41.

Set the points vertical to each other and the origin.

42. Select the **Dimension** tool from the ribbon.

43. Set the distance between the two points to **20 mm.**
Set the distance from the top edge and the top point to **15 mm**.
Click the Green check to exit the sketch.

44. Select the **Hole** tool from the ribbon.

45.
Select **Simple** Hole.
Set the Termination to **Through**.
Set the Standard to **ISO**.
Set the Hole type to **Drilled**.
Set the Drill size to **4 mm**.
Select the two points.
Verify that Part 1 is shown.
Click the Green check.

46. Select the **Fillet** tool.

47. Select the two bottom angled edges for a 10 mm radius fillet.
Click the Green check.

48. Select the **Chamfer** tool.

49.

Hold down the right mouse to rotate the model.

Select the top horizontal edges for a 5 mm equal distance chamfer.
Click the Green check.

50.

Right click on the part in the parts browser.
Select **Rename**.

51.

Type **Angled Plate**.
Click ENTER.

52.

Right click on the Part Studio tab and select Rename.

53.

Type **Angled Plate**.
Click ENTER.

54.

Right click on the part in the Parts browser and select **Assign material.**

55.

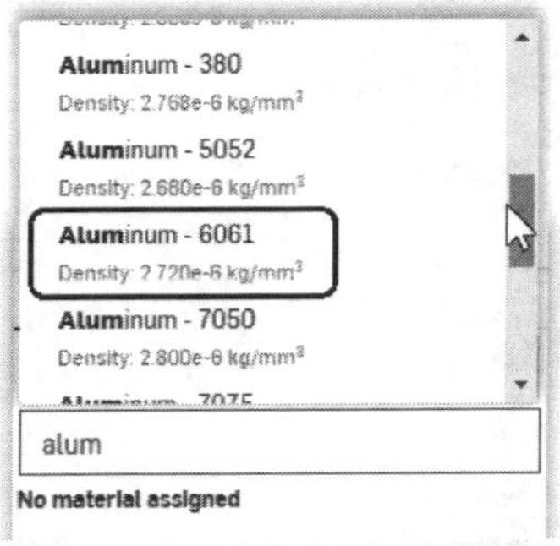

Type **alum** in the search field to list all the aluminum materials.
Scroll down to select **Aluminum – 6061.**

56.

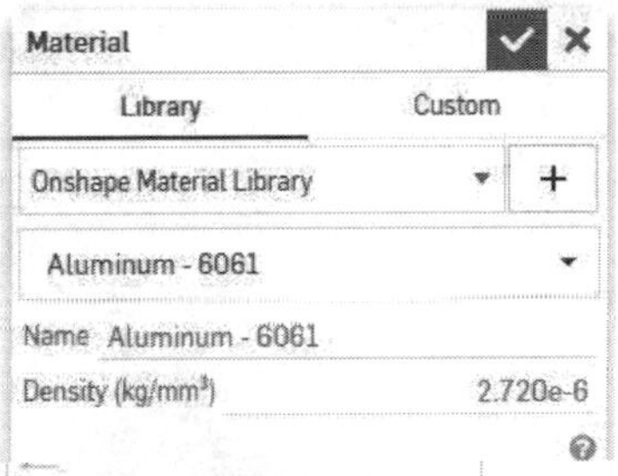

Click the Green check.

57.

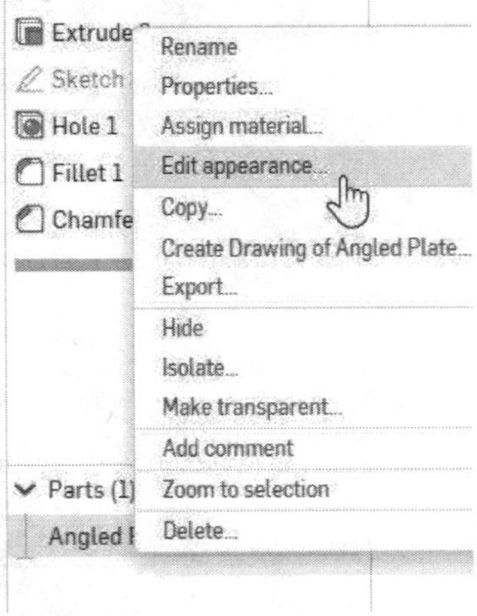

Right click on the part in the Parts browser and select **Edit appearance.**

58.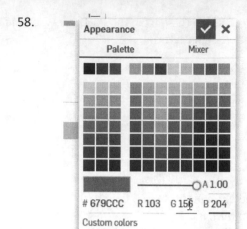

Enter R **103.**
Enter G **156.**
Enter B **204.**
Click the Green check.

59.

Close the document.

Chapter 4: Base Plate Project

Estimated Time: 45 minutes
Objectives:

- Extrude
- Hole
- Import DWG
- Show/Hide Sketch
- Linear Pattern
- Chamfer
- Edit Appearance

This part will be used to create drawings and explore the different drawing tools.

1. 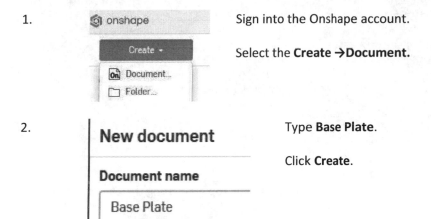 Sign into the Onshape account.

   Select the **Create →Document.**

2. Type **Base Plate**.

   Click **Create**.

3. Set the **Workspace units**.

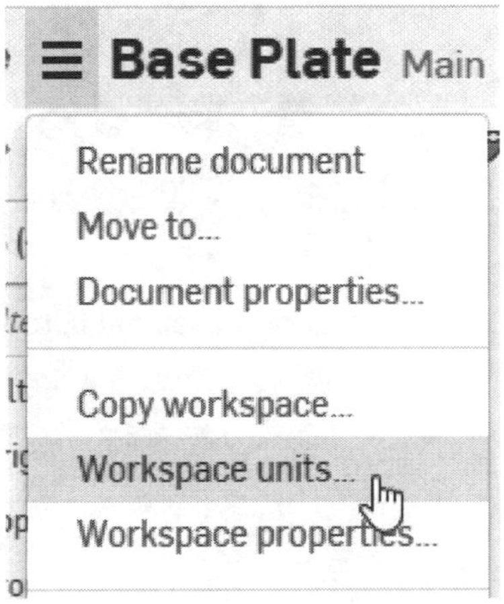

4.

Workspace units ✓ ✕

Default length unit
Millimeter ▼

Length display decimals
0.1 ▼

Default angle unit
Degree ▼

Angle display decimals
0 ▼

Default mass unit
Kilogram ▼

Mass display decimals
0.1 ▼ ❓

Set the Default Length unit to **Millimeter**.
Set the Length display decimals to **0.1**
Set the Angle display decimals to **0**.
Set the Default mass unit to **Kilogram**.

Green check.

5.

✓ Default geometry
⊙ Origin
🔲 Top
🔲 Front   Rename
🔲 Right   ✏ New sketch...
   🔲 Offset plane
   Hide

Highlight the **Top** plane in the browser.

Right click and select **New Sketch**.

6.

Switch to a **Top** view using the Orientation Cube.

7.

▭ ∨ ⊙ ∨ ⌒ ∨ ⬠ ∨

□ Corner rectangle   G
▢ Center point rectangle   R

Select the **Center point rectangle** tool.

8. 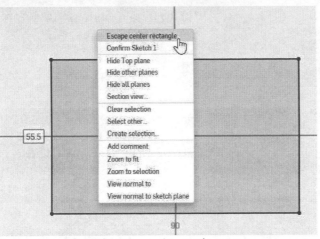 Draw the rectangle using the origin as the center point.

Right click and select **Escape center rectangle.**

9. 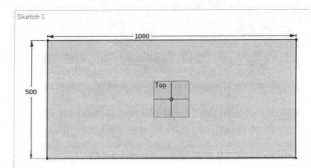 Select the **Dimension** tool.

10. Add dimensions to define the rectangle as 500 mm x 1060 mm.

11. Select the **Extrude** tool.

12. Extrude the rectangle **30 mm**.

Green check.

13.

Isometric
Dimetric
Trimetric

Use the Display Options to switch to an **Isometric** view.

14.

Select the top face of the extrude.

Right click and select **New Sketch**.

15.

Switch to a **TOP** view.

16.

Sketch 2

Sketch plane
Face of Extrude 1

☐ Disable imprinting
☑ Show constraints
☑ Show overdefined

Verify that **Show Constraints** is enabled for the Sketch.

17.

Select the **POINT** tool.

18.

Place four points – one at each corner.

Right click and **select Escape point.**

Escape point
Confirm Sketch 2

19.

Add horizontal and vertical constraints to align the points.

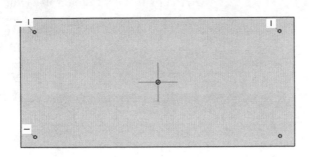

Hint: You need to escape after two points are selected and then restart the Constraint command.

Once the constraints are placed, the points should adjust their position, so they are aligned.

20.   Select the **Dimension** tool.

21.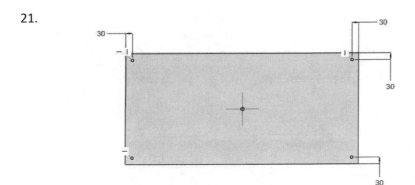

Position each point so it is 30 mm from the edge in the vertical and horizontal direction.

22.   Sketch 2   Click the **Green Check** to close the sketch.

Sketch plane
Face of Extrude 1

☑ Show constraints

☑ Show overdefined

23.   Select the **Hole** tool.

24.

Hole 1
Simple
Through
Standard ISO
Hole type Drilled
Drill size 20
20 mm
Sketch points to place holes
Vertices of Sketch 2
Merge scope
Part 1

Set the Hole to **Simple**.
Set the Distance to **Through**.
Set the Drill size to **20 mm.**
Select the sketch with the points from the browser.
Green check.

25.

On the right side of the display window, there is a set of tabs.

Click on the **Hole table** tab.

You see a table with the holes you just placed.

Part 1		
Tag	Size	Qty
A	Ø 20 mm THRU	4

26.

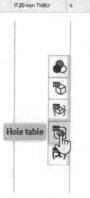

Click on the tab to collapse the panel.

27.

Select the top face of the extrude.

Right click and select **New Sketch**.

28.

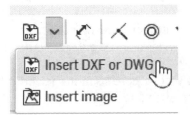

Select **Insert DXF or DWG**.

29.

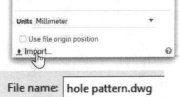

At the bottom of the panel,
Set the Units to **Millimeter**.
Click **Import**.

30.

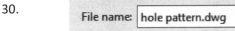

Locate the *hole pattern.dwg* file in the Chapter 04 folder.
Click **Open**.

31.

There will be a pause while the file is imported.
The file will be listed.
Click on the file.

32.

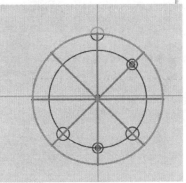

Select the outer circle and the lines.

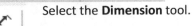

Select the **Reference** tool.

This converts those elements to reference or construction elements.

33.

Select the **Dimension** tool.

34.

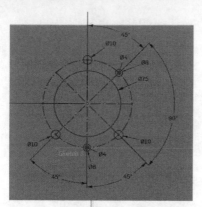

Add diameter dimensions to the circles and angular dimensions between the construction lines.

35.

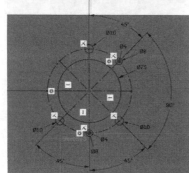

Add a coincident constraint between the center of the circles and the outer construction circle.

Add a vertical constraint to the vertical construction line.

Add a horizontal constraint to the horizontal construction line.

Add a coincident constraint between the end points of the construction lines and the outer construction circle.

Add a concentric constraint between the outer construction circle and the inner large circle.

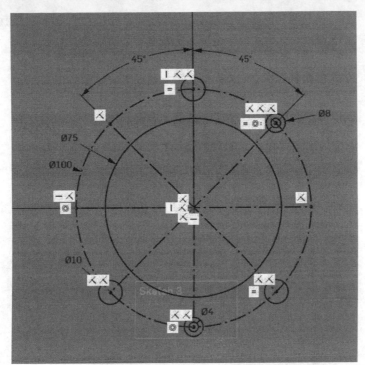

The sketch should be fully constrained when you are done adding dimensions and constraints.

36. 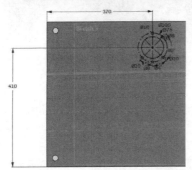 Position the hole pattern 410 mm above the lower left corner and 370 mm to the right of the lower left corner.

37.  Green check to exit the sketch.

38. Select the **HOLE** tool.

39.

Select the center point of the hole pattern.
Set the Hole to Simple.
Set the Standard to **Custom**.

Type **75 mm** for the hole diameter.

Click Green Check.

40.

Highlight the hole pattern sketch in the browser.

Right click and select **Show**.

41. Select the **HOLE** tool.

42.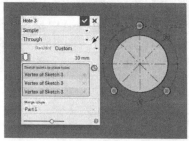

Select the three centerpoints for the 10 mm circles.

Set the Hole to **Simple**.
Set the Standard to **Custom**.
Set the Distance to **Through**.
Set the Diameter to **10 mm**.
Green check.

43. Select the **HOLE** tool.

44.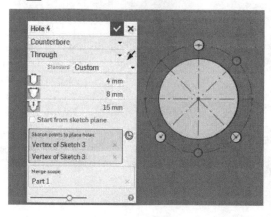

Set the Hole type to **Counterbore**.
Set the Distance to **Through**.
Set the Drill diameter to **4 mm**.
Set the Counterbore diameter to **8 mm**.
Set the Counterbore depth to **15 mm**.
Select the two counterbore centerpoints in the hole pattern.
Green check.

45.  Click to **Hide** the hole pattern sketch.

46. 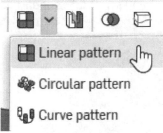 Use the Display Options to switch to an **Isometric** view.

47. 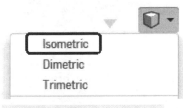 Select **Linear pattern**.

48. 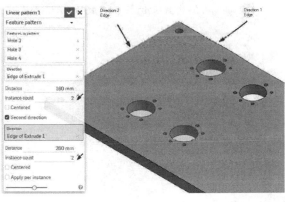 Select **Feature pattern**.
Select the holes placed for the hole pattern.
Select the top edge of the plate.
Set the Distance to **160 mm.**
Set the Instance Count to **2.**
Enable **Second Direction**.
Select the left edge of the plate.
Set the Distance to **260 mm**.
Set the Instance Count to **2**.

Green Check.

49. Select the **CHAMFER** tool.

50.

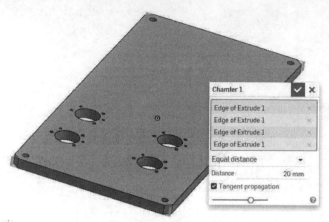

Select the four corners of the plate.
Set the Chamfer to **Equal distance**.
Set the Distance to **20 mm**.
Green Check.

51.

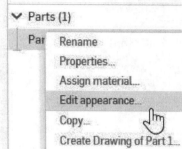

Highlight the part in the browser.

Right click and select **Edit appearance**.

52.

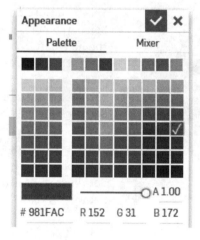

Assign Color Code **#981FAC** to the plate.

Green Check.

53.

onshape

Close the document.

Part Studio: Loop Hanger

Estimated Time: 60 minutes
Objectives:
- Create Surface
- Convert to Sheet Metal
- Create Hole
- Flanges
- Variable Studio
- Configurations

1.

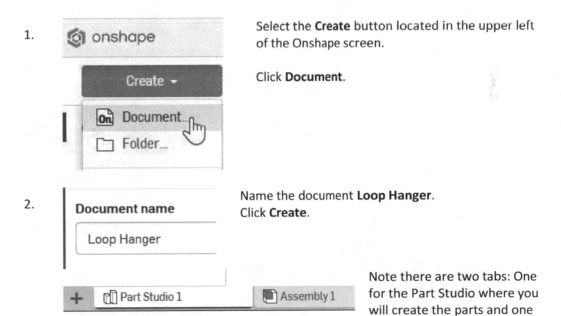

Select the **Create** button located in the upper left of the Onshape screen.

Click **Document**.

2.

Name the document **Loop Hanger**.
Click **Create**.

Note there are two tabs: One for the Part Studio where you will create the parts and one for Assembly.

3.

Click the + tab.

Select **Create Variable Studio**.

Variable Studio is used to create table-driven parts.

These are parts that use the same geometry with different dimensions or features.

4.

Type in the values as shown in the table.

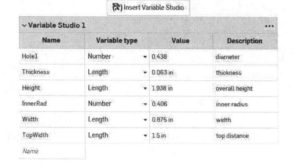

Name	Variable type	Value	Description
Hole1	Number	0.438	diameter
Thickness	Length	0.063 in	thickness
Height	Length	1.938 in	overall height
InnerRad	Number	0.406	inner radius
Width	Length	0.875 in	width
TopWidth	Length	1.5 in	top distance
Name			

5.

Switch to the **Part Studio 1** tab.
Rename **Loop Hanger**.

6.

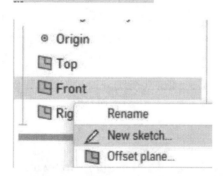

Place a New sketch on the **Front** plane.

7.

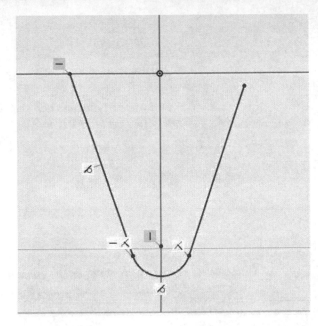

Place two angled lines and an arc.

The right angled line's top endpoint is slightly below the left angled line's top endpoint.

Add a vertical constraint between the origin and the arc center.

Add a tangent constraint between the arc and the two angled lines.

Add a horizontal constraint between the lower endpoints of the angled lines.
I add the inner radius dimension to the arc.

8.

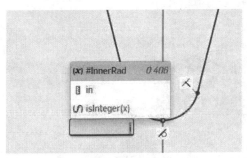

Select the dimension tool. Erase the existing dimension and start typing the desired variable name.

Select **#InnerRad**.
Click ENTER.
Place a point at the lower quadrant of the arc.

9.

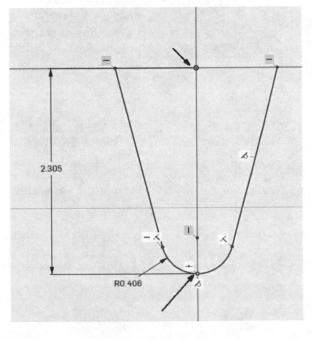

Add a vertical dimension between the origin and the lower quadrant.

10.

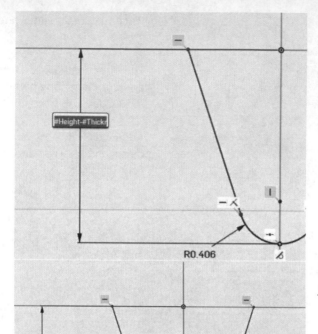

Click Backspace to delete the dimension value.

Type **#Height-#Thickness**.

As you type, you can select the two variable names.

Onshape does the calculation for you and displays the value.

11.

Add a vertical dimension between the origin and the top right endpoint of the angled line.
Set it to **#Thickness*2**.

12.

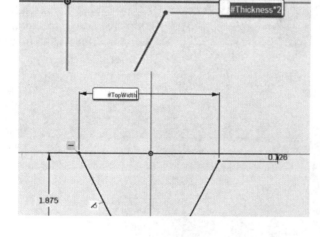

Add a horizontal dimension between the two line endpoints.

Set it to the **TopWidth** variable.

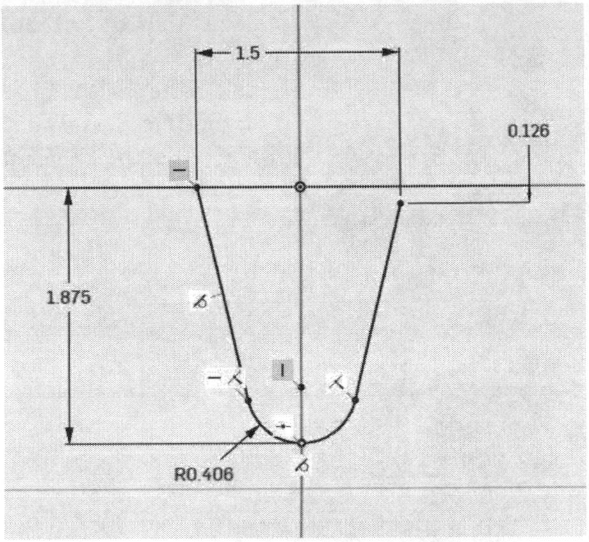

The sketch should look like this.

1.5

0.126

1.875

R0.406

13.

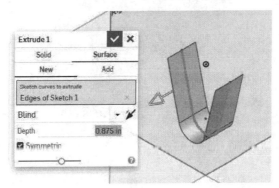

Select the **Extrude** tool.

Enable **Surface**.
Select the two lines and the arc.
Enable **Symmetric**.
In the Depth field, delete the displayed value.
Type **W** and select **#Width**.

Green Check.

14.

Select the **Sheet Metal Model** tool.

15.

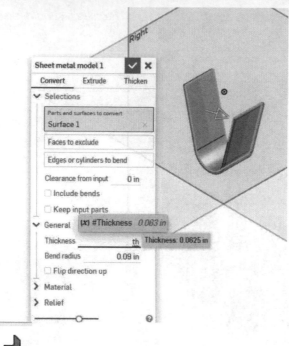

Select the **Convert** tab.

Select the surface to convert. In the Thickness field, type **TH** and select the **#Thickness** variable.

Green check.

16.

Select the **Flange** tool.

17.

Select the left edge of the model.
Set the Flange alignment to **Inner**.
Set the Distance to **#TopWidth**.
Set the Bend angle to **103.25 deg**.

Green check.

18.

Select the **Flange** tool.

19.

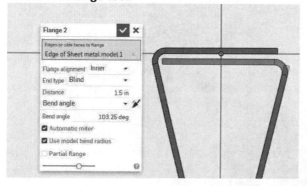

Select the right edge of the model.
Set the Flange alignment to **Outer**.
Set the Distance to **#TopWidth**.
Set the Bend angle to **103.25 deg**.

Green check.

Our model so far.

20.

Select the top face for a **New Sketch**.

21.

Place a point at the origin.

Exit the sketch.

22.

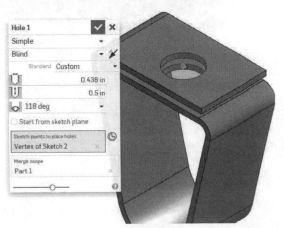

Select the **Hole** tool.
Set the Hole Type to **Simple**.
Set the Termination to **Blind**.
Set the Standard to **Custom**.
Assign the **#Hole1** value to the diameter.
Set the Depth to **0.5 in**.
Select the point as the location for the hole.
Green check.

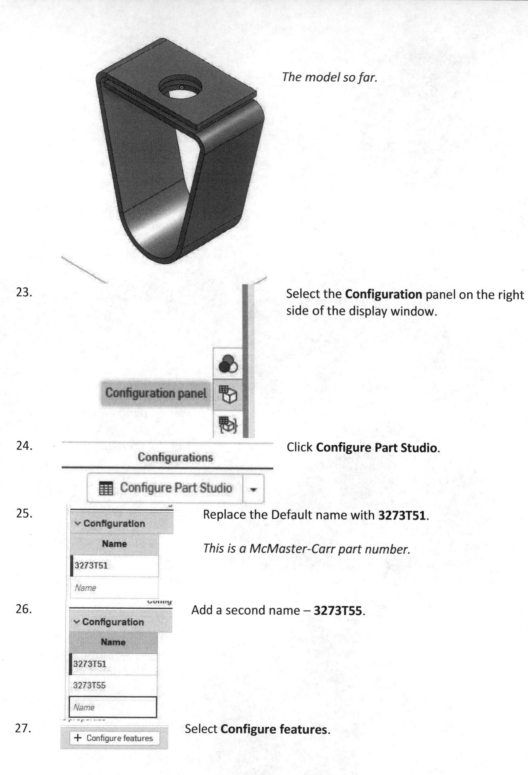

The model so far.

23.   Select the **Configuration** panel on the right side of the display window.

24.   Click **Configure Part Studio**.

25.   Replace the Default name with **3273T51**.

This is a McMaster-Carr part number.

26.   Add a second name – **3273T55**.

27.   Select **Configure features**.

28. 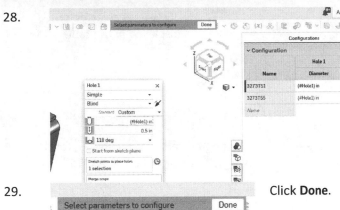 Select the Hole 1 in the browser list.

Click on the hole diameter field.

Note that the Hole1 variable is copied over to the configuration table.

29. 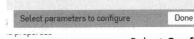 Click **Done**.

30. Select **Configure features**.

31. Left click on **Sketch 1** in the browser to make it visible.

32. 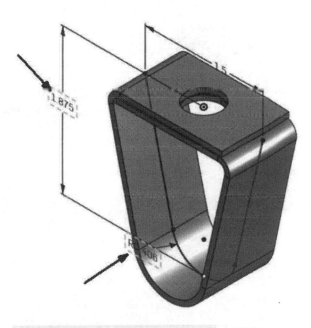 Select the 1.875 dimension and the R.406 dimension.

33. 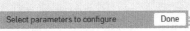 Click **Done**.

34.

		Sketch 1		Hole 1
Name	Radius	Distance	Diameter	
3273T51	(#InnerRad) in	#Height-#Thickness	(#Hole1) in	
3273T55	(#InnerRad) in	#Height-#Thickness	(#Hole1) in	

The values selected are displayed in the table.

35.

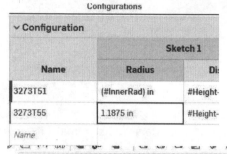

Change the value for the Radius for the 3273T55 configuration to **1.1875 in**.

36.

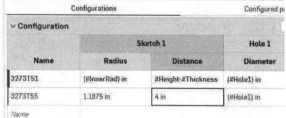

Change the value for the Distance for the 3273T55 configuration to **4 in**.

37.

Highlight the **3273T55** configuration. Right click and select **Switch to 3273T55**.

The upper flanges fail.

38.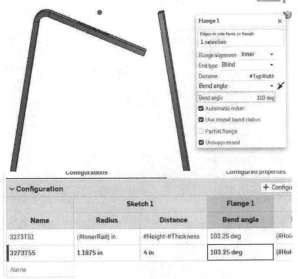
+ Configure features

Select **Configure features**.

39.
📄 Extrude 1
📄 Sheet metal model 1
📄 Flange 1
📄 Flange 2
✏️ Sketch 2

Select **Flange 1**.

40. Click on the **Bend angle**.

Flange 1   ✕

Edges or side faces to flange
1 selection

Flange alignment   Inner   ▼
End type   Blind   ▼
Distance   #TopWidth
Bend angle   ▼ ✏️
Bend angle   103 deg
☑ Automatic miter.
☑ Use model bend radius
☐ Partial flange
☑ Unsuppressed

The Bend angle for Flange 1 is now included in the table.

Configurations   Configured properties

∨ Configuration   + Configu

	Sketch 1		Flange 1	
Name	Radius	Distance	Bend angle	
3273T51	(#InnerRad) in	#Height-#Thickness	103.25 deg	(#Hol⋅
3273T55	1.1875 in	4 in	103.25 deg	(#Hol⋅
Name				

41.

Configuration	Sketch 1		Flange 1
Name	Radius	Distance	Bend angle
3273T51	(#InnerRad) in	#Height-#Thickness	103.25 deg
3273T55	1.1875 in	4 in	81 deg
Name			

Change the value for the Bend angle for the **3273T55** configuration to **81 deg**.

42.

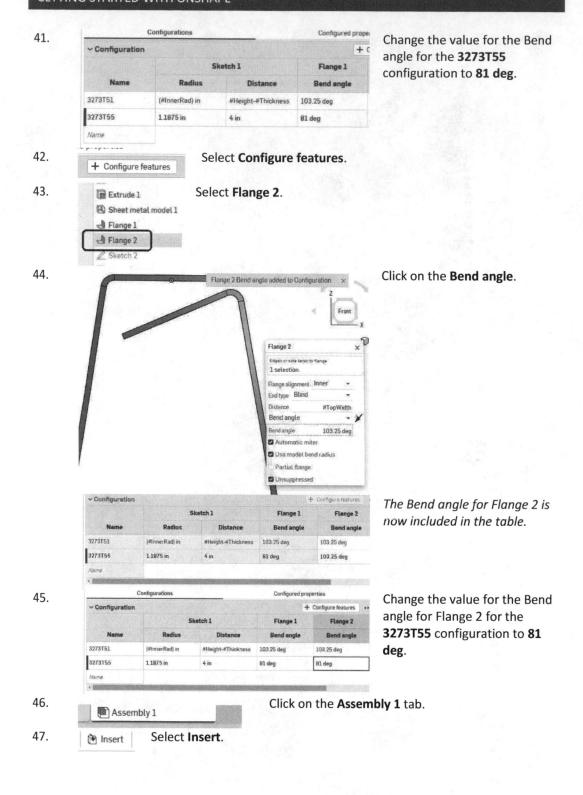

+ Configure features

Select **Configure features**.

43.

📷 Extrude 1
📷 Sheet metal model 1
📷 Flange 1
📷 Flange 2
📷 Sketch 2

Select **Flange 2**.

44.

Flange 2 Bend angle added to Configuration

Flange 2
Edges or side faces to flange
1 selection
Flange alignment Inner
End type Blind
Distance #TopWidth
Bend angle
Bend angle 103.25 deg
☑ Automatic miter
☑ Use model bend radius
☐ Partial flange
☑ Unsuppressed

Click on the **Bend angle**.

Configuration	Sketch 1		Flange 1	Flange 2
Name	Radius	Distance	Bend angle	Bend angle
3273T51	(#InnerRad) in	#Height-#Thickness	103.25 deg	103.25 deg
3273T55	1.1875 in	4 in	81 deg	103.25 deg
Name				

The Bend angle for Flange 2 is now included in the table.

45.

Configurations		Configured properties		
Configuration	Sketch 1		Flange 1	Flange 2
Name	Radius	Distance	Bend angle	Bend angle
3273T51	(#InnerRad) in	#Height-#Thickness	103.25 deg	103.25 deg
3273T55	1.1875 in	4 in	81 deg	81 deg
Name				

Change the value for the Bend angle for Flange 2 for the **3273T55** configuration to **81 deg**.

46.

📷 Assembly 1

Click on the **Assembly 1** tab.

47.

📷 Insert

Select **Insert**.

48.

Note that both configurations are listed in the drop down list.

Select **3272T51** to be inserted.

Click Part 1.

Click inside the display window to place.

49.

Select **3272T55** to be inserted.

Click Part 1.

Click inside the display window to place.

Two Part 1 components are listed in the assembly even though they are different configurations and different part numbers.

50.

Select the **Loop Hanger** tab.

51.

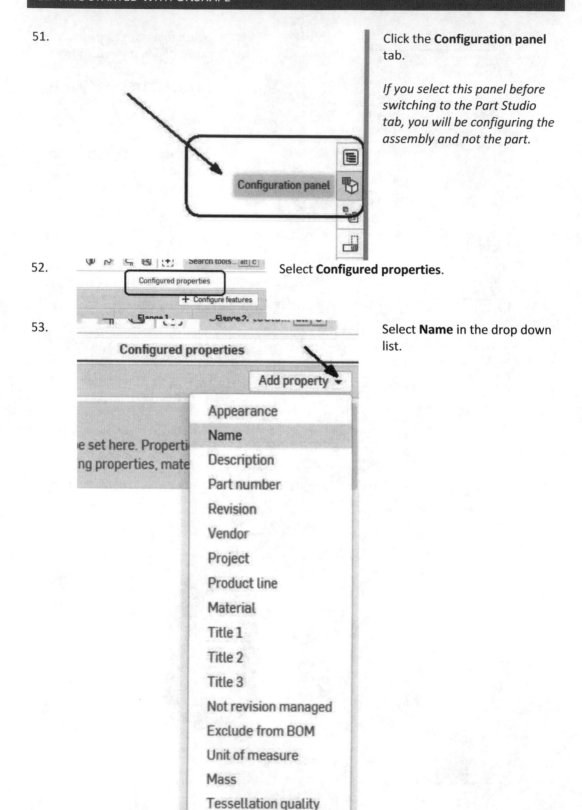

Click the **Configuration panel** tab.

If you select this panel before switching to the Part Studio tab, you will be configuring the assembly and not the part.

52.

Select **Configured properties**.

53.

Select **Name** in the drop down list.

54.

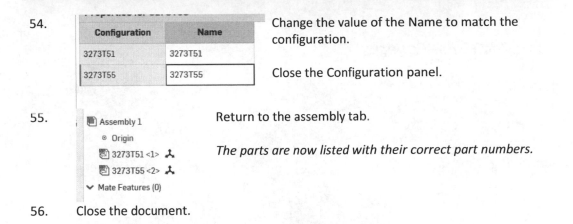

Configuration	Name
3273T51	3273T51
3273T55	3273T55

Change the value of the Name to match the configuration.

Close the Configuration panel.

55.

Assembly 1
  Origin
  3273T51 <1>
  3273T55 <2>
  Mate Features (0)

Return to the assembly tab.

The parts are now listed with their correct part numbers.

56.   Close the document.

Extra: Explain how configurations are different from using the Variable Studio? What are the advantages of using the Variable Studio? What are the advantages of using configurations?

Extra: Additional Projects

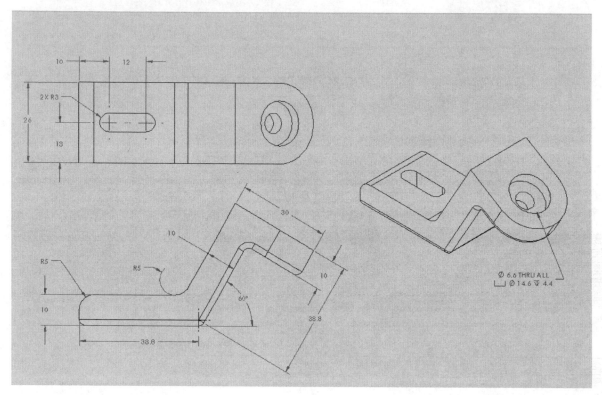

EX4-1 LEVER
UNITS ARE IN MILLIMETERS

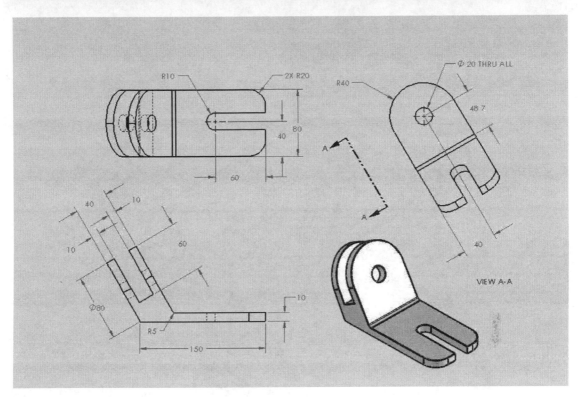

EX4-2 JIG
UNITS ARE IN MILLIMETERS

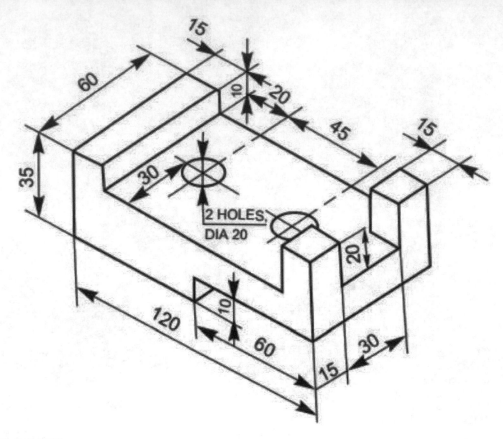

EX4-3 BLOCK
UNITS ARE IN MILLIMETERS

Chapter 5: Scooter Project

Part One: The Deck

Estimated Time: 2 hours
Objectives:

- Set Units for a Document
- Rename Tabs
- Rename Sketches
- Rename Features
- Create Hole
- Using Construction Lines
- Extrude
- Fillet
- Loft
- Edit Appearance

1.
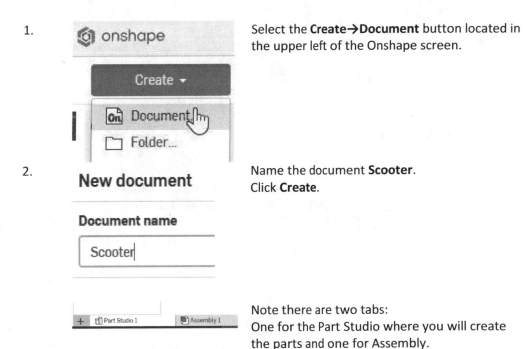
Select the **Create→Document** button located in the upper left of the Onshape screen.

2.
Name the document **Scooter**.
Click **Create**.

Note there are two tabs:
One for the Part Studio where you will create the parts and one for Assembly.

3.

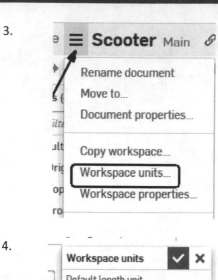

Select the drop down under the Documents menu.

Select **Workspace Units**.

4.

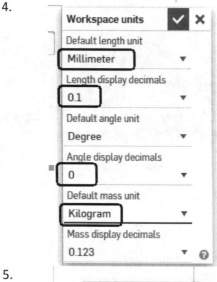

Set the Length to **Millimeter.**
Set the Length display decimals to **0.1.**
Set the Angle display decimals to **0.**
Set the Mass to **Kilogram**.

Click the Green check.

5.

Select the Part Studio 1 tab.
Right click and select **Rename.**

6.

Type **Deck**.
Click ENTER.

7.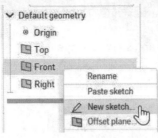

Select the **Front plane** in the browser.
Right click and select **New sketch**.

8.

Enable **Show constraints.**

9.

Use the Orientation cube to change to a Front view.

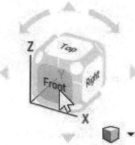

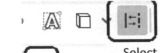

We will create the sketch shown.

Hints: I used the vertical construction line to mirror one side of the sketch.

10.

Enable **Construction**.

11.

Select the **LINE** tool.

12.

Create a vertical line starting at the origin.

Right click and select **Escape line**.

13.

Verify that **Construction** is no longer enabled.

14.

Select the **LINE** tool.

15.

Use the LINE tool to place the lines shown.
Don't worry about the dimensions yet.

We will leverage the symmetry of the profile.

Right click and select **Escape line** when you have
completed the figure.

16.

Select the vertical construction line.

17.

Select the **Mirror** tool.

18.

Disable **Construction**.
It became enabled when the vertical construction line
was selected.

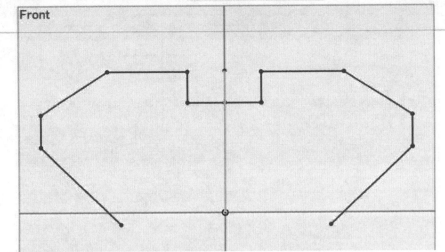

19. Select the lines – they will be mirrored as they are selected.
You can window around the lines or use a crossing to select.

20.  Right click and select **Escape mirror**.

21. Select the **3 point arc** tool.

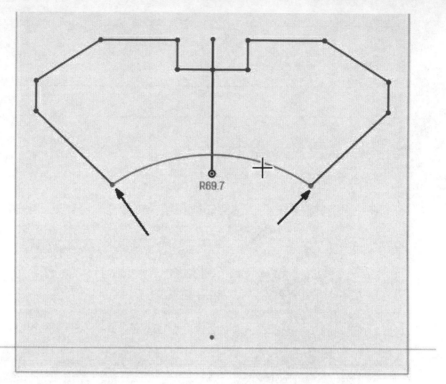

22. Place the arc to close the polygon. Left click on the left endpoint first.
Then left click on the right endpoint.
Then left click to place the arc.

23. 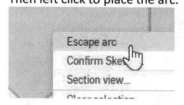 Right click and select **Escape arc**.

24. Select the **Dimension** tool.

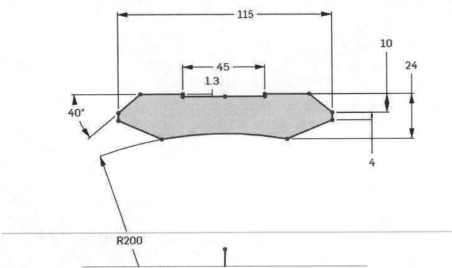

25.   Add the dimensions.
       Notice that the sketch is under-defined.

26.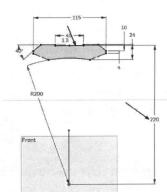

       Add a vertical dimension between the lower
       horizontal line and the origin.

       Set the value to **220 mm**.

27.   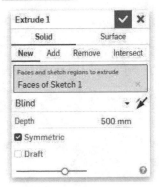   Select the **Extrude** tool from the ribbon.

28.   Enable **Symmetric**.
       Set the value to **500 mm**.
       Click the **Green check** button.

29.

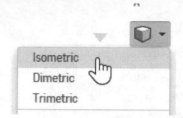

Use the Display Options cube to switch to an isometric view, so you can inspect your part.

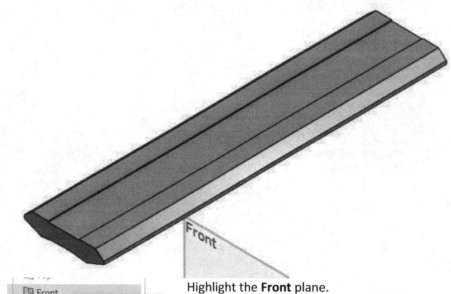

30.

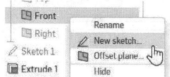

Highlight the **Front** plane.
Right click and select **New sketch**.

31.

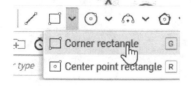

Use the Orientation cube to change to a Front view.

32.

Select the Corner rectangle.

33.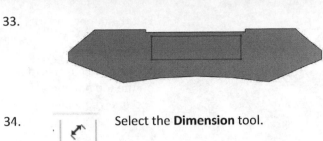

Draw a rectangle below the upper notch.

Right click and select **Escape rectangle**.

34. 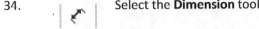   Select the **Dimension** tool.

35.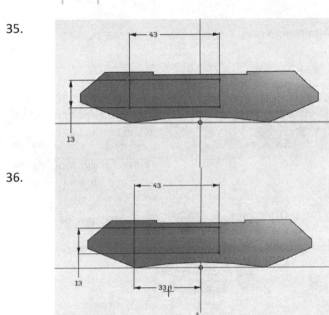

Add a **43 mm** horizontal dimension and a **13 mm** vertical dimension to resize the rectangle.

36.

Place a horizontal dimension between the origin and left vertical side of the rectangle.

37.

Type **43/2** to center the rectangle on the model.

38.

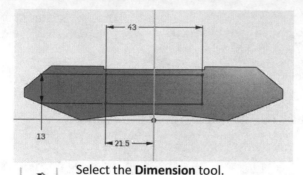

Exit the dimension tool.

Move the rectangle below the top notch by left clicking on the top horizontal line and dragging down.

39. Select the **Dimension** tool.

40.

Place a **2 mm** vertical dimension from the lower edge of the notch to the top of the rectangle.

The rectangle should show as fully defined.

41. Select the **Extrude** tool.

42.

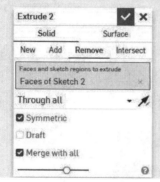

Left click on **Remove**.
Enable **Symmetric**.
Set the depth to **Through All**.
Enable **Merge with all**.
Click the **Green check** button.

43.

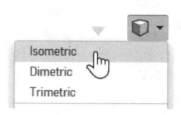

Rename the Extrude **Center Cut**.

44. Use the Display Options cube to switch to an isometric view, so you can inspect your part.

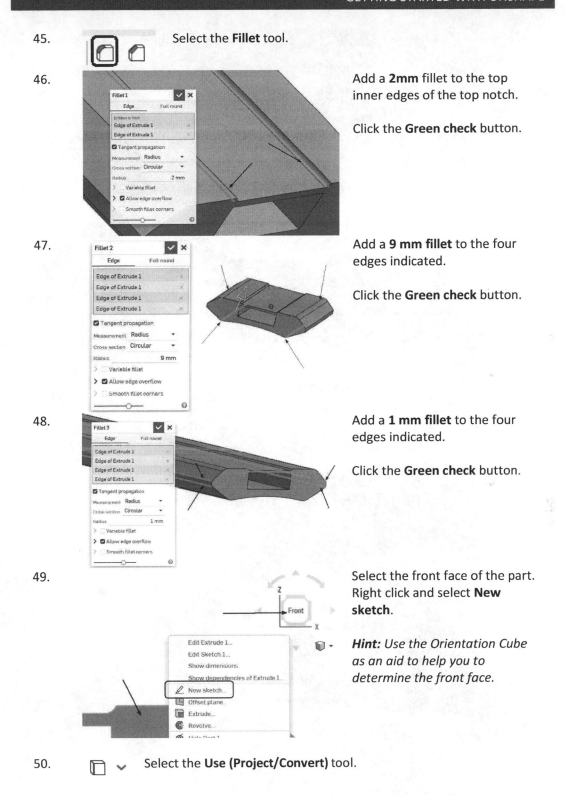

45. Select the **Fillet** tool.

46. Add a **2mm** fillet to the top inner edges of the top notch.

    Click the **Green check** button.

47. Add a **9 mm fillet** to the four edges indicated.

    Click the **Green check** button.

48. Add a **1 mm fillet** to the four edges indicated.

    Click the **Green check** button.

49. Select the front face of the part. Right click and select **New sketch**.

    Hint: *Use the Orientation Cube as an aid to help you to determine the front face.*

50. Select the **Use (Project/Convert)** tool.

51.

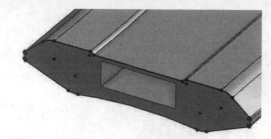

Select the front face of the part.

The outer edges are copied into the sketch.

You may need to select each element or a few elements at a time for the projection to work.

52. 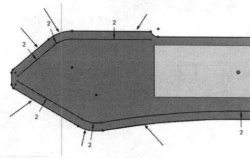 Select the **Offset** tool.

53.

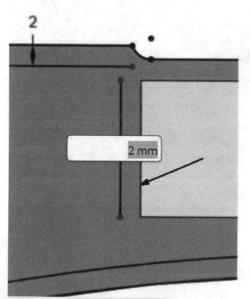

Select the lines and arcs on the left side as indicated by the arrows.

Click on the arrow to set the offset into the part.

Change the dimension to **2 mm**.

54. Select the **Offset** tool.

55.

Offset the vertical edge **2 mm** from the inner rectangle.

2

2 mm

56.  Select the **Fillet Arc** tool from the ribbon.

Sketch Fillet (Shift-f)

57.

Add a fillet arc with a **2 mm** radius.

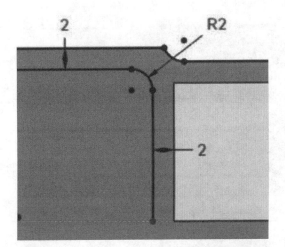

58.

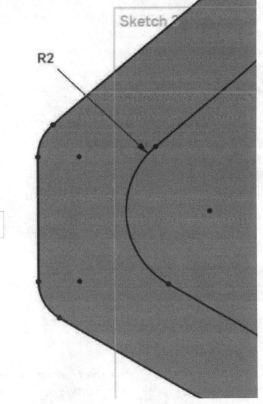

Add a fillet arc with a **2 mm** radius to the left side of the polygon.

59.

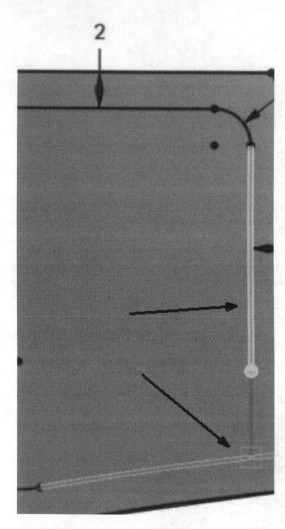

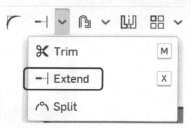

Select the **Extend** tool from the ribbon.

Extend the right vertical line until it intersects with the lower offset arc.

60.

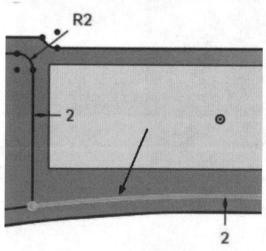

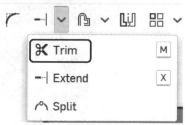

Use the **Trim** tool to remove the right portion of the offset arc.

GETTING STARTED WITH ONSHAPE

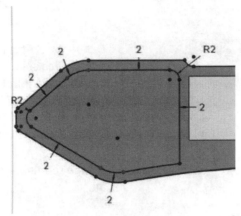

The closed polygon should appear as shown.

Next, we will mirror the polygon to the other side of the face.

61.

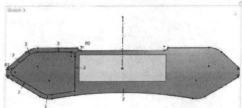

Draw a vertical construction line starting at the origin.

62.   Select the **Mirror** tool.

63.

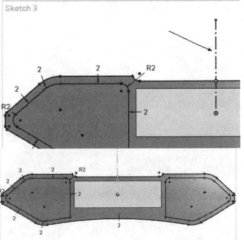

Select the vertical construction line.

64.

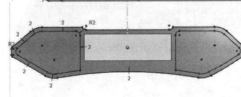

Window around the sketched polygon to select.

65.   Right click and select **Escape Mirror**.

Escape mirror

Confirm Sketch 3

Copy sketch

66. Select the **Extrude** tool from the ribbon.

67. | Solid | | Surface | |
    |---|---|---|---|
    | New | Add | **Remove** | Intersect |

    Select the **Remove** option.

68. Delete the sketch as selected and instead select two interior points inside the sketch.

69. Set the termination to **Through all**.

    Enable **Merge with all**. Click the **green check** to complete.

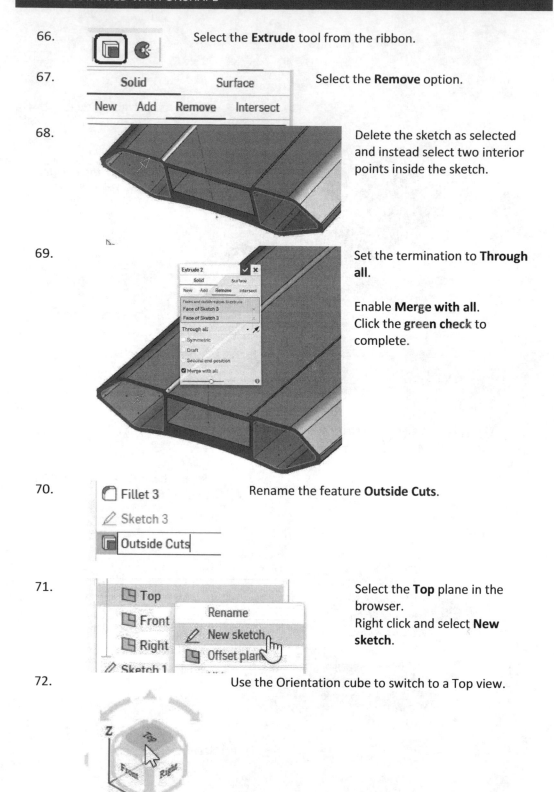

70. 🗋 Fillet 3

    ✏ Sketch 3

    Outside Cuts

    Rename the feature **Outside Cuts**.

71. Select the **Top** plane in the browser.
    Right click and select **New sketch**.

72. Use the Orientation cube to switch to a Top view.

73. 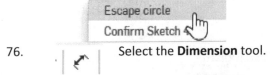 Select the **Circle** tool.

74. 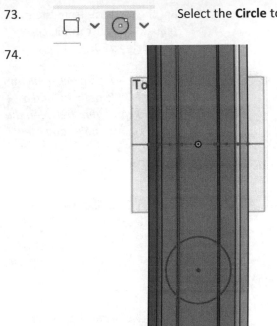 Place a circle centered on the top of the part but not coincident to the origin.

75. Right click and select **Escape circle**.

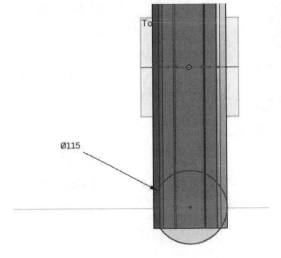

76. Select the **Dimension** tool.

77. Set the diameter of the circle to **115 mm**.

Ø115

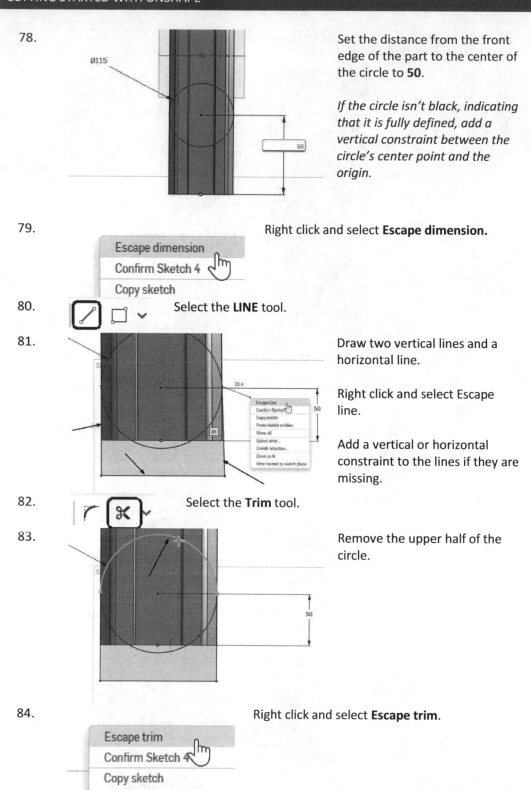

78. Set the distance from the front edge of the part to the center of the circle to **50**.

If the circle isn't black, indicating that it is fully defined, add a vertical constraint between the circle's center point and the origin.

79. Right click and select **Escape dimension.**

80. Select the **LINE** tool.

81. Draw two vertical lines and a horizontal line.

Right click and select Escape line.

Add a vertical or horizontal constraint to the lines if they are missing.

82. Select the **Trim** tool.

83. Remove the upper half of the circle.

84. Right click and select **Escape trim**.

85. 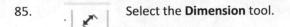 Select the **Dimension** tool.

86.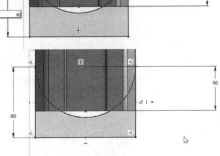

Add an 80 mm dimension to the vertical line.

87.

Use geometric constraints to complete the sketch definition.

Set the two vertical lines equal.

88. Use the Display Options cube to switch to an isometric view, so you can inspect your part.

Isometric

Dimetric

Trimetric

89. Select the **Extrude** tool from the ribbon.

90.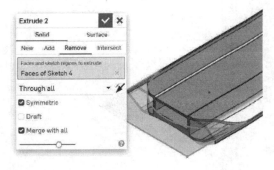

Select the **Remove** option.
Enable **Symmetric**.
Set the termination to **Through all**.
Enable **Merge with all**.
Click the green check.

If you don't see a clean cut, edit the sketch and ensure the vertical lines are coincident to the outside edges of the deck and the endpoints of the arc are coincident to the vertical lines.

91.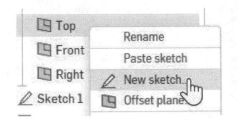

Select the **Top** plane in the browser.
Right click and select **New Sketch**.

92.

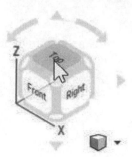

Use the Orientation cube to change to the **Top** view.

93.

Use the arrows on the Orientation Cube to rotate view.

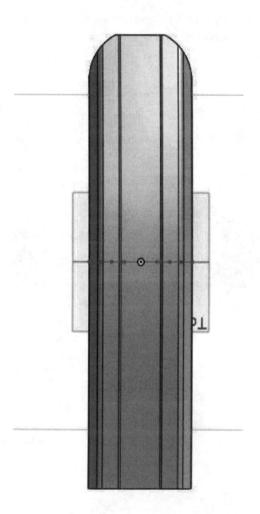

The top view of the deck should appear as shown.

94. Select the **LINE** tool.

95.

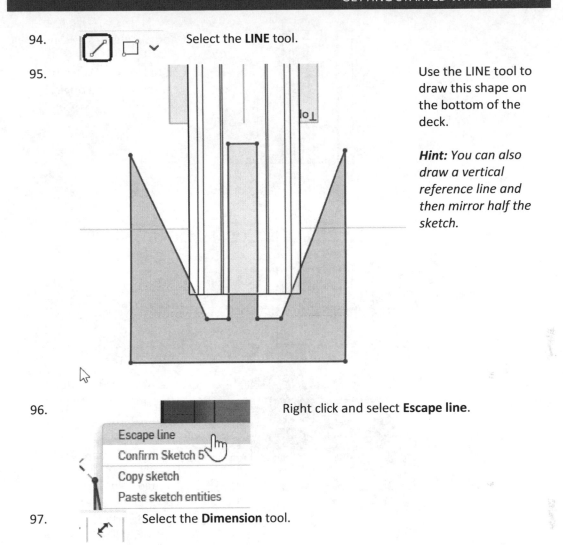

Use the LINE tool to draw this shape on the bottom of the deck.

Hint: You can also draw a vertical reference line and then mirror half the sketch.

96. Right click and select **Escape line**.

> Escape line
> Confirm Sketch 5
> Copy sketch
> Paste sketch entities

97. Select the **Dimension** tool.

98.

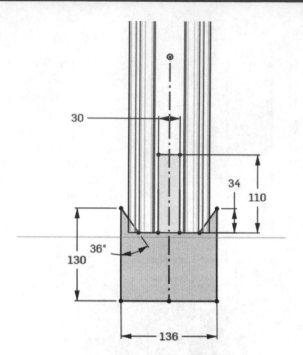

Add dimensions as shown.

Note that the sketch is centered on the origin.

99.

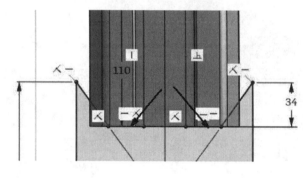

The two upper horizontal lines are aligned to the edge.

Use geometric constraints to fully define the sketch. I added coincident constraints to align the outside vertical and small horizontal lines with the outer edges of the part. *Onshape doesn't have a collinear sketch constraint – use coincident or horizontal instead!*

100.

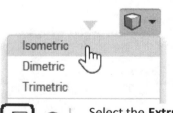

Isometric

Dimetric

Trimetric

Use the Display Options cube to switch to an isometric view, so you can inspect your part.

101.

Select the **Extrude** tool from the ribbon.

102.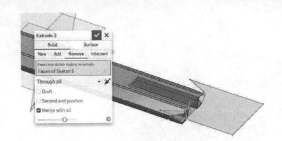

Select the **Remove** option.
Enable **Symmetric**.
Set the termination to **Through all**.
Enable **Merge with all**.

Click the green check.

103. ✏ Sketch 5

Rename the feature **Rear Cut.**

 ▣ Rear Cut

104.

Select the **Fillet** tool from the Features ribbon.

105.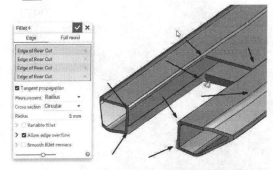

Select the inside edges of the rectangular opening.

Set the Radius to **5 mm**.

Click the Green check to complete.

106.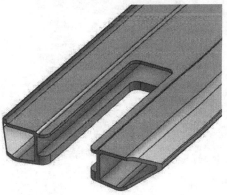

Inspect the part to see that the fillets were properly applied.

107.

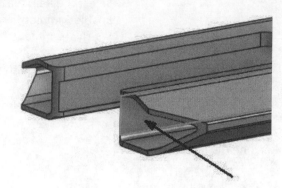

Select the inside face of the bottom area of the deck. Right click and select **New sketch**.

108. 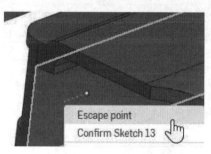 Select the **Point** tool from the Sketch ribbon.

109.

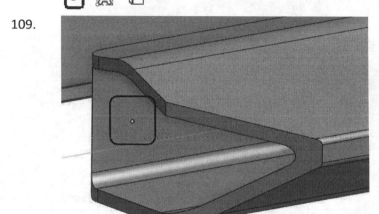

Place the point on the face.

110.

Right click and select **Escape point**.

Escape point
Confirm Sketch 13

111.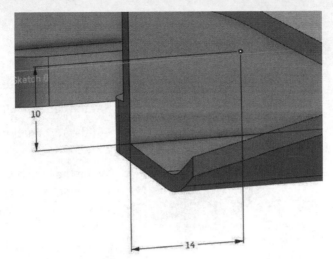

Add dimensions to locate the point 10 mm above the inside edge and 14 mm from the outer edge.

112.

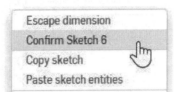

Right click in the display window. Select **Confirm Sketch <#>**.
Note: Your sketch number might be different from mine.

113. Select the **Hole** tool on the feature ribbon.

114.

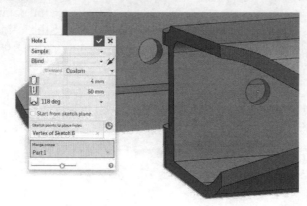

Set the Hole to be **Simple.**

Set the diameter to **4 mm**.

Set the depth to **50 mm**.

Set the Termination to **Blind**.

When the blue box indicating sketch points is highlighted, select the point.

Left click in the Merge scope box.

Select the deck part.

Press the **Green Check**.

115.

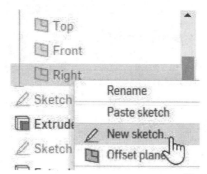

Select the **Right** plane in the browser. Right click and select **New Sketch**.

116.

Use the Orientation cube to change to the **Left** view.

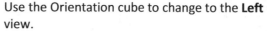

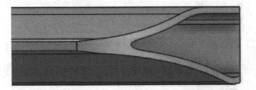

We are going to be adding a sketch profile to this end of the deck – the end without the hole.

117.

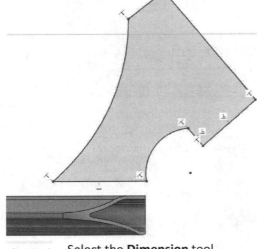

Draw the shape shown using two 3 point arcs and five lines.

118. 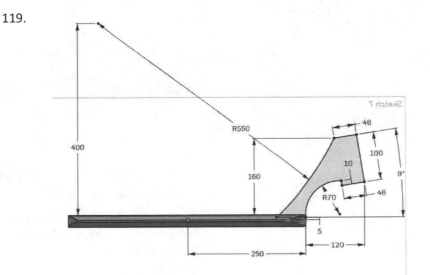 Select the **Dimension** tool.

119.

Add dimensions as shown.

Hint: *You can add dimensions from the reference planes to the desired points on the sketch.*

120.

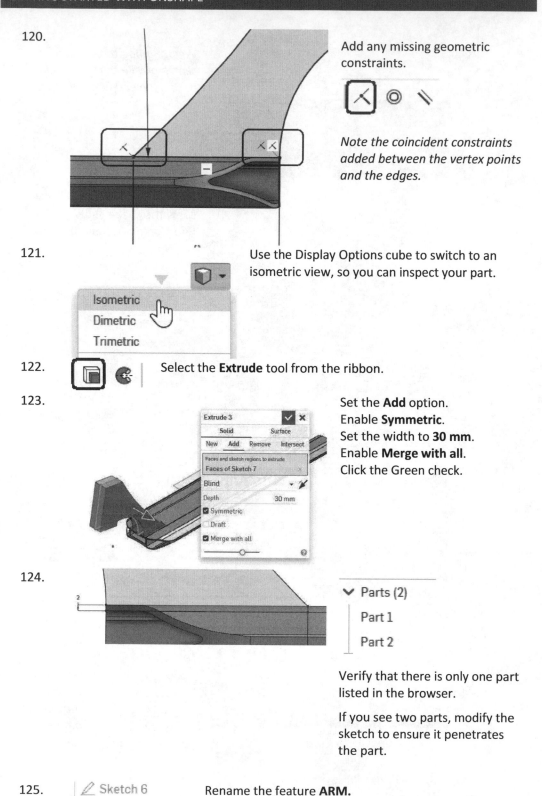

Add any missing geometric constraints.

Note the coincident constraints added between the vertex points and the edges.

121.

Isometric

Dimetric

Trimetric

Use the Display Options cube to switch to an isometric view, so you can inspect your part.

122.

Select the **Extrude** tool from the ribbon.

123.

Extrude 3

Solid   Surface

New   **Add**   Remove   Intersect

Faces and sketch regions to extrude
Faces of Sketch 7

Blind

Depth   30 mm

☑ Symmetric

☐ Draft

☑ Merge with all

Set the **Add** option.
Enable **Symmetric**.
Set the width to **30 mm**.
Enable **Merge with all**.
Click the Green check.

124.

∨ Parts (2)

Part 1

Part 2

Verify that there is only one part listed in the browser.

If you see two parts, modify the sketch to ensure it penetrates the part.

125.

Sketch 6

Arm

Rename the feature **ARM.**

126.

Select the bottom face of the upper Arm for a new sketch.

Edit Arm...
Edit Sketch 7...
Show dimensions
Show dependencies of Arm...
New sketch...
Offset plane...

127.

Orient the display to a **Bottom** view.

Bottom

128.

Select the **3 point circle** tool.

Center point circle c
3 point circle
Ellipse

or type

129.

Select the upper and lower corners of the face.

130.

Right click and select **Escape 3 point circle.**

Escape 3 point circle
Confirm Sketch 14

131.

Select the **EXTRUDE** tool.

132.

Enable **Add.**
Select **Up to Face.**
Switch the direction of the extrude.
Select the top face.
Enable **Merge with all.**
Green check.

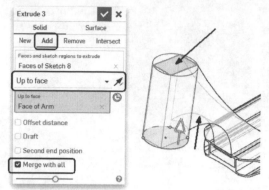

133.

Rename the extrude **Fork Cylinder.**

134.

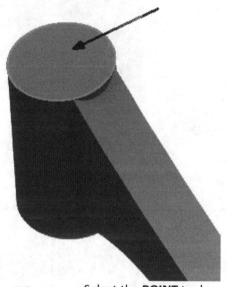

Select the top face for a new sketch.

135.

Select the **POINT** tool.

136.

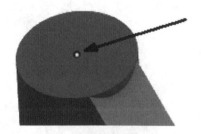

Place a point concentric to the cylinder.

Exit the Sketch.

137.

Select the **Hole** tool.

138.

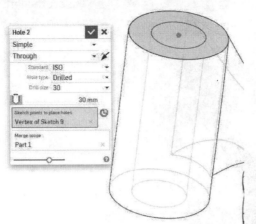

Set the Hole as **Simple**.
Set the Depth as **Through**.
Set the Standard as **ISO**.
Set the Hole type as **Drilled**.
Set the Drill size as **30 mm**.
Select the point.
Click Green check.

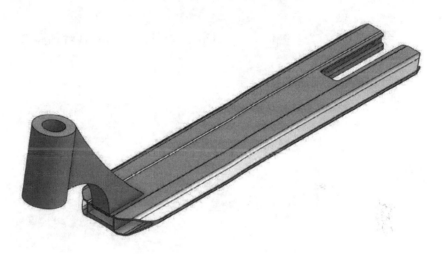

Inspect your part to see how far you have come!

139.      Select the **Sketch** tool on the ribbon.

140.   Left click in the Sketch
                                             Plane text box.

141.

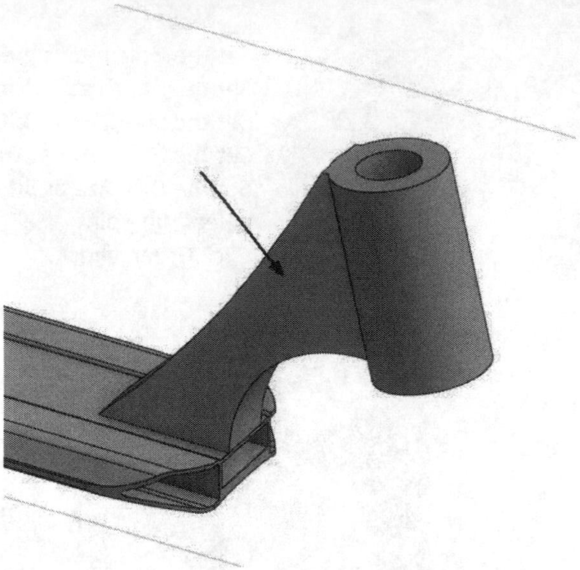

Select the left side of the arm.

Hint: Use the Orientation Cube to help you figure out which is the left side.

142.

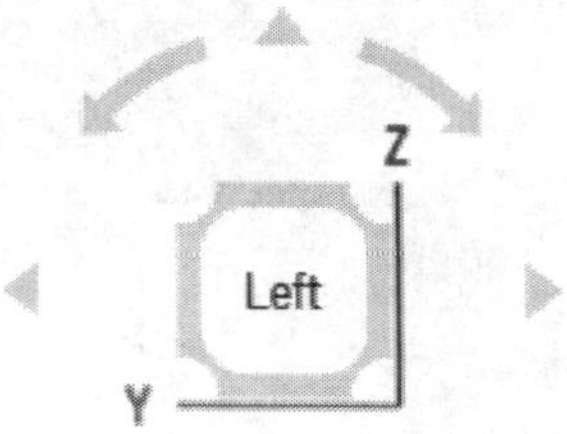

Orient the view to the **LEFT** view.

143.

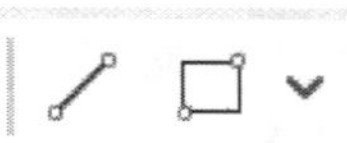

Select the **LINE** tool.

144.

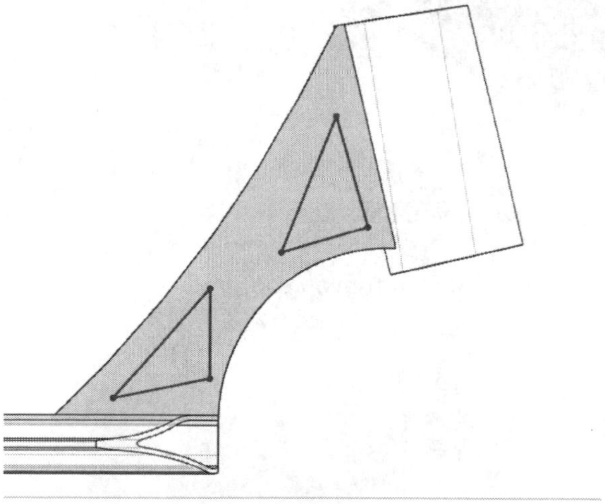

Draw two triangles.

145.

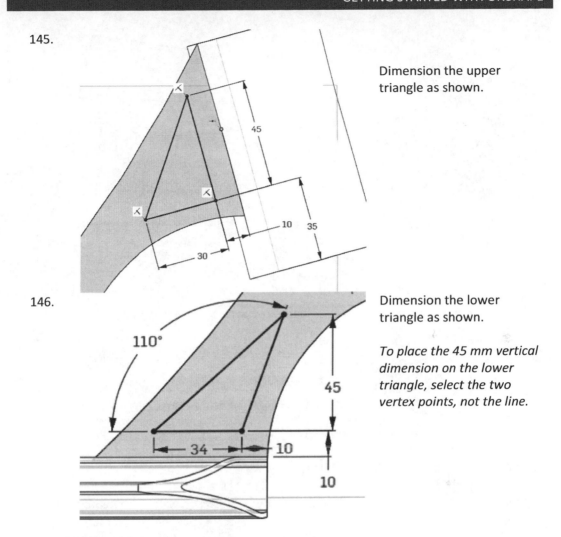

Dimension the upper triangle as shown.

146.

Dimension the lower triangle as shown.

To place the 45 mm vertical dimension on the lower triangle, select the two vertex points, not the line.

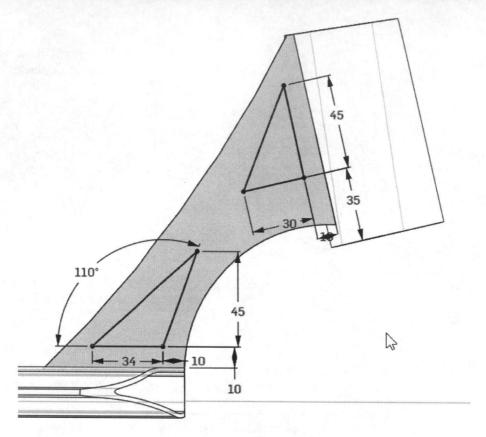

147. Select the **Extrude** tool.

148. Select the **Remove** option.
Enable **Symmetric**.
Set the Termination to
Through all.
Enable **Merge with all**.
Select the **Green check**.

Extrude 3

Solid   Surface

New   Add   Remove   Intersect

Faces and sketch regions to extrude
Faces of Sketch 10

Through all

☑ Symmetric
☐ Draft
☑ Merge with all

149. Select the **Fillet** tool.

150.

Add a **5 mm** fillet to the six inside edges of the triangular cuts.

Select the **Green check**.

151.

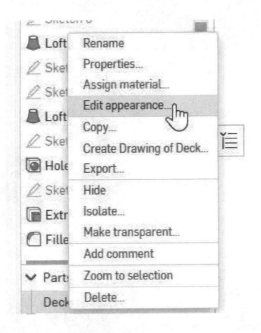

Rename the Part in the browser to **Deck**.

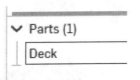

152.

Select the Deck in the browser.
Right click and select **Edit Appearance**.

153.

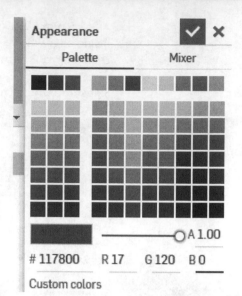

Set the R value to **17.**
Set the G value to **120**.
Set the B value to **0.**
Click the Green check.

154.

Close Onshape.

Part Two: The Brake

Estimated Time: 90 minutes
Objectives:

- Extrude
- Sheet Metal Part
- Tab
- Flange
- Mirror
- Slot
- Hole
- Fillet
- Assign Appearance
- Rename Part
- Finish Sheet Metal Part
- Flat Pattern

1.

   Log in to your Onshape account.
   Left click on the **Scooter** document.

2.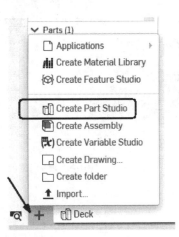

   Select the + symbol at the bottom left of the screen.
   Select **Create Part Studio**.

3.

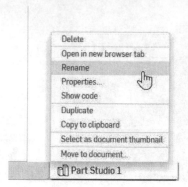

A new Part Studio tab will be added.
Right click on the tab.
Select **Rename.**

4.

Type **Brake**.

```
🗐 Brake|
```

5.

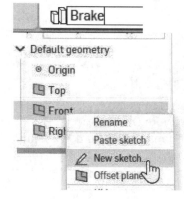

Select the **Front** plane in the browser.
Right click and select **New sketch**.

We are going to create this sketch.

6.

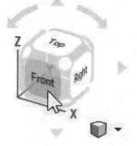

Use the Orientation Cube to switch to a **Front** view.

7.   ⊙ ∨   Select the **Centerpoint Circle** tool.

8. 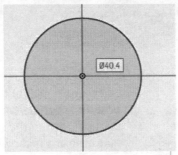 Place a centerpoint circle coincident to the origin.

9. Right click and select **Escape circle.**

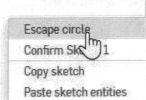

10. Select the **LINE** tool.

11. 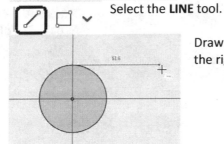 Draw a horizontal line from the top of the circle to the right.

12. Right click and select **Escape line**.

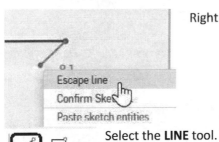

13. Select the **LINE** tool.

14. 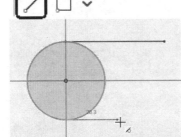 Draw a shorter horizontal line from the bottom of the circle to the right.

15. 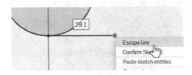 Right click and select **Escape line**.

16.   Select the **3 point arc** tool.

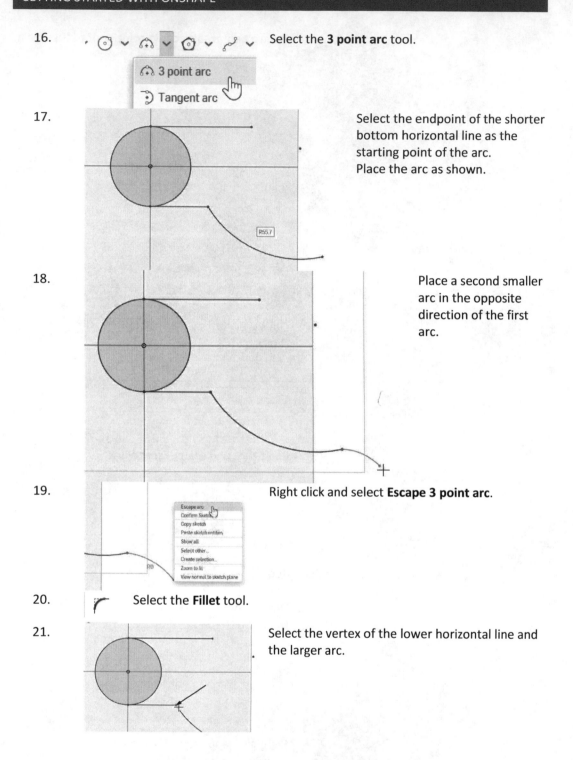

17.   Select the endpoint of the shorter bottom horizontal line as the starting point of the arc.
Place the arc as shown.

18.   Place a second smaller arc in the opposite direction of the first arc.

19.   Right click and select **Escape 3 point arc**.

20.   Select the **Fillet** tool.

21.   Select the vertex of the lower horizontal line and the larger arc.

22. Set the fillet radius to **5 mm**.

23. Right click and select **Escape fillet**.

24. Select the **Trim** tool.

25. Select the right side of the circle to be deleted.

26. Right click and select **Escape trim**.

27. Select the **TANGENT** constraint.

28. Select the two arcs.

29. Right click and select **Escape create tangent constraint**.

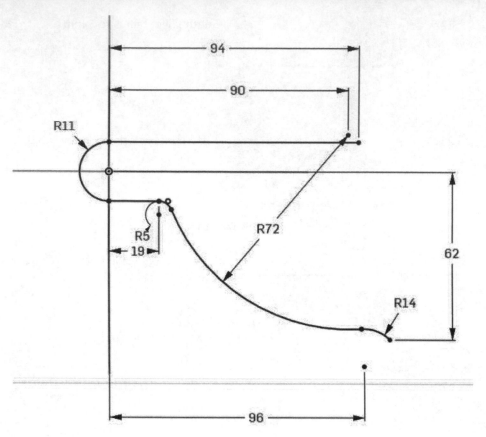

30.   Add dimensions.

31.   Switch to an **Isometric** view.

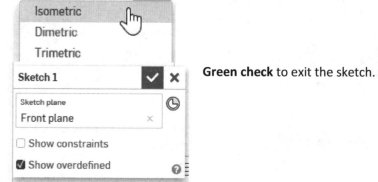

32.   **Green check** to exit the sketch.

33.   Select the **Sheet Metal model** tool from the ribbon.

This converts the part to a sheet metal model.

34.

Enable **Extrude**.
Select the elements in the sketch.

Set the Depth to **16 mm**.
Set the Thickness to **2 mm**.

Accept the defaults for all the other values.

Green check.

35.

∨ Default geometry

⊙ Origin

▢ Top

▢ Front

▢ Right

✎ Sketch 1

▣ Sheet metal model 1

Hide the sketch in the browser.

Notice that the browser lists a sheet metal model.

36.

Select the **top face** of the brake for a new sketch.

37.

□ ∨

Select the **Corner Rectangle** tool.

38.

Draw a corner rectangle with the top right vertex coincident with the top face.

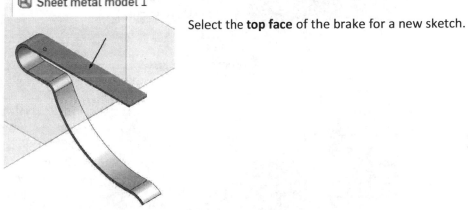

39.

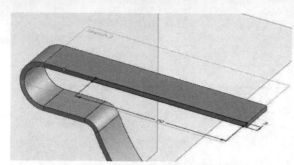

Add dimensions so that the rectangle is 70 mm long and 3 mm high.

40.

Right click and select **Confirm Sketch 2**.

41. 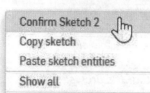 Select the **Tab** tool from the sheet metal features.

42.

Select **Sketch 2** for the Tab profile.
Select the top face of the brake as the flange to merge.
Set the Subtraction offset to **0 mm**.
Click Green check.

43. Select the **Flange** tool.

44.

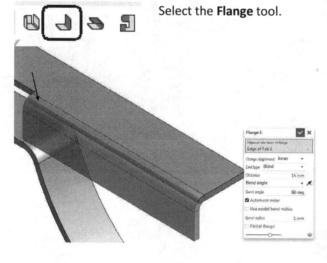

Select the top edge of the tab.
Set the Flange alignment to **Inner**.
Set the End type to **Blind**.
Set the Distance to **14 mm**.
Click the Bend angle arrow to switch the direction of the flange to down.
Disable **Use model bend radius.**
Set the Bend radius to **2 mm.**

Green check.

45.

Sheet metal model 1

Sketch 2

Tab 1

Flange 1

Check in the browser to review the feature names.

46.

Select the **Mirror** tool from the ribbon.

47.

Mirror 1

Part mirror

New   Add   Remove   Intersect

Entities to mirror
Part 1

Mirror plane

Select **Part 1** from the browser's lower panel.

48.

Mirror 1

Part mirror

New   Add   Remove   Intersect

Entities to mirror
Part 1

Mirror plane

Left click in the *Mirror plane* field.

49.

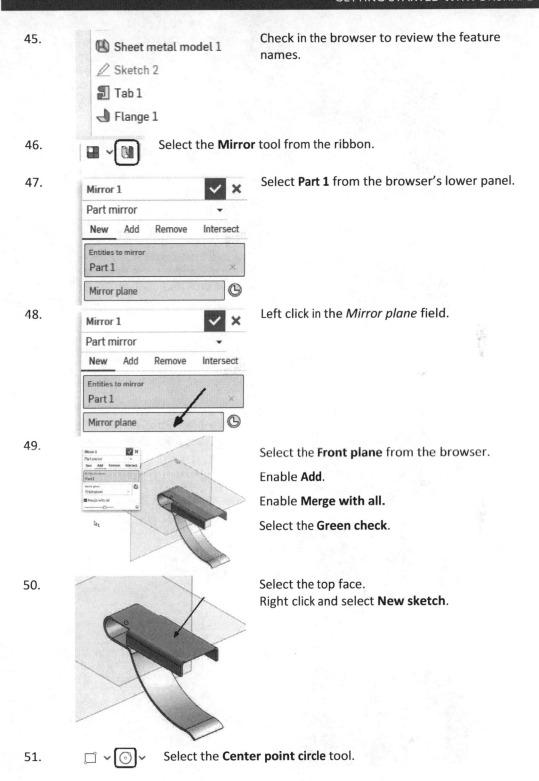

Select the **Front plane** from the browser.

Enable **Add**.

Enable **Merge with all.**

Select the **Green check**.

50.

Select the top face.
Right click and select **New sketch**.

51.

Select the **Center point circle** tool.

52. Place a circle on the face.

53. Right click and select **Escape center point circle**.

Escape circle
Confirm Sketch
Copy sketch

54. Add a horizontal constraint between the circle's center point and the origin.

55. Select the **Line** tool.

56. Place two horizontal lines at the upper and lower quadrants.
You can use object tracking to line up the endpoint of the lines with the circle's center.

57. Draw a short vertical line to connect the two lines.

58. Select the **Trim** tool.

59. Delete the inside arc.

60. 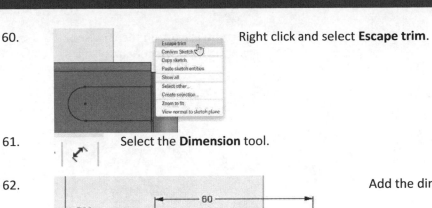 Right click and select **Escape trim**.

61. Select the **Dimension** tool.

62. Add the dimensions.

63. Right click and select **Escape dimension**.

Escape dimension
Confirm Sketch
Copy sketch

64. Add tangent constraints between the horizontal lines and the arc to fully define the sketch.

65. Select the **Extrude** tool.

66. 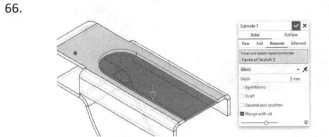 Enable **Remove**.
Set the Termination to **Blind.**
Set the distance to **2 mm.**
Enable **Merge with all.**
Select the **Green check**.

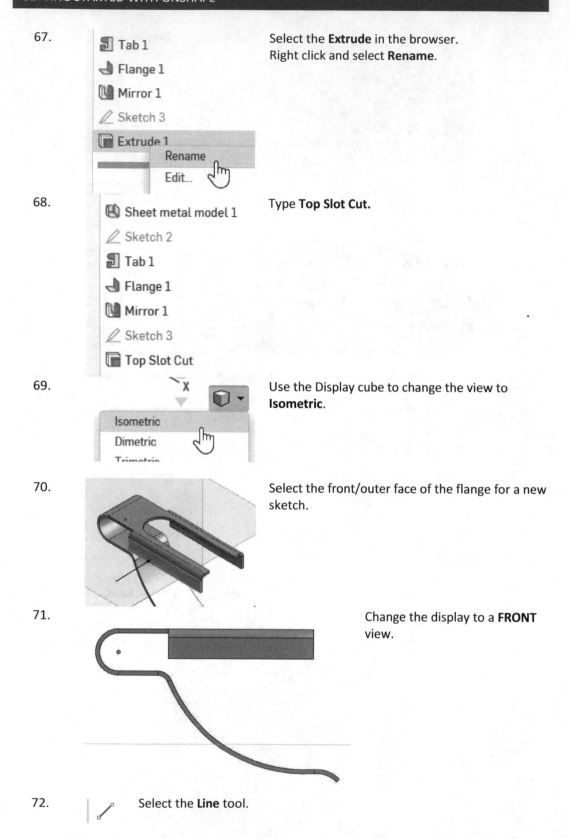

67.
Select the **Extrude** in the browser.
Right click and select **Rename**.

68.
Type **Top Slot Cut.**

69.
Use the Display cube to change the view to **Isometric**.

70.
Select the front/outer face of the flange for a new sketch.

71.
Change the display to a **FRONT** view.

72.
Select the **Line** tool.

73.

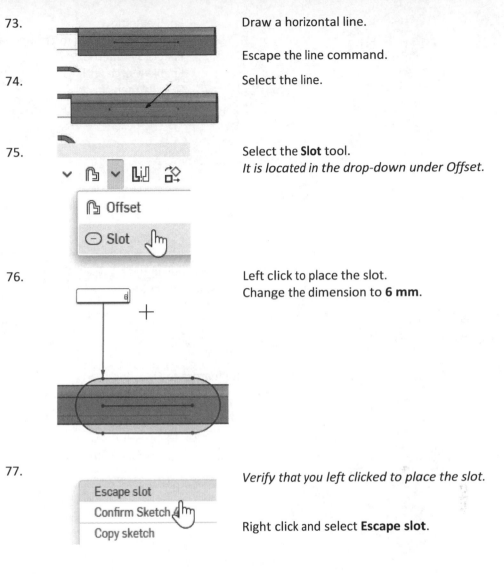

Draw a horizontal line.

Escape the line command.

74.

Select the line.

75.

Select the **Slot** tool.
It is located in the drop-down under Offset.

76.

Left click to place the slot.
Change the dimension to **6 mm**.

77.

Verify that you left clicked to place the slot.

Right click and select **Escape slot**.

78.

Select the **Dimension** tool.

79.

Add a length dimension of **20 mm.**

Position the slot **24 mm** from the right edge and **5 mm** above the bottom edge of the flange.

80.

Select the **Extrude** tool.

81.

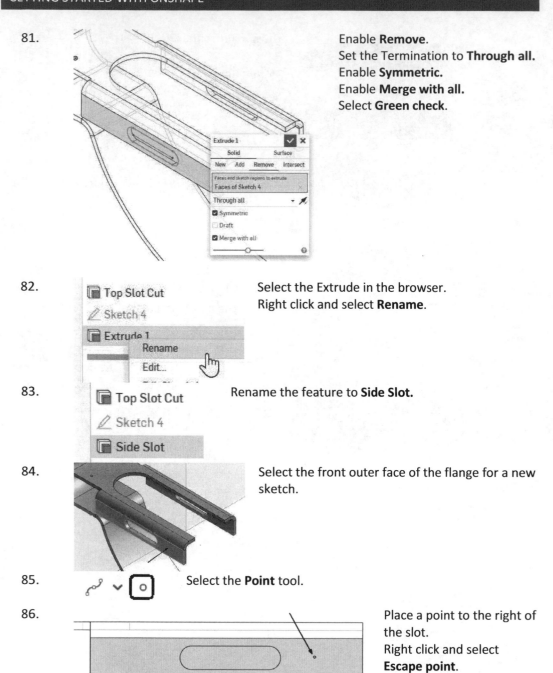

Enable **Remove**.
Set the Termination to **Through all.**
Enable **Symmetric.**
Enable **Merge with all.**
Select **Green check**.

82.

Select the Extrude in the browser.
Right click and select **Rename**.

83.

Rename the feature to **Side Slot.**

84.

Select the front outer face of the flange for a new sketch.

85.

Select the **Point** tool.

86.

Place a point to the right of the slot.
Right click and select **Escape point**.

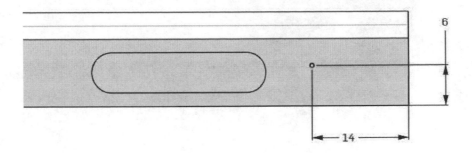

87. Use the dimension tool to locate the point **14 mm** from the right side of the flange and **6 mm** from the bottom edge of the flange.

   Exit the sketch.

88. Select the **Hole** tool.

89. 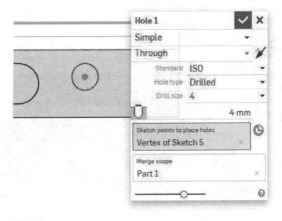 Set the Hole to be **Simple**.
   Set the Standard to **ISO.**
   Set the Hole diameter to **4 mm**.
   Set the Hole to be **Through**.
   Select the point.
   Select the **Green check**.

90. Select the **Fillet** tool.

91.

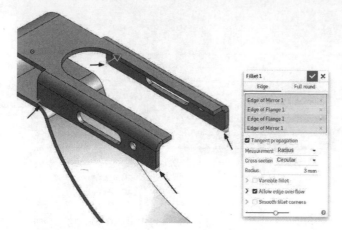

Set the fillet radius to **3 mm**.

Select the four bottom corners of the two flanges.

Select the **Green check**.

92.

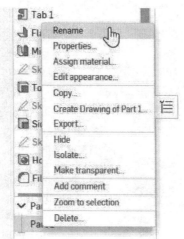

Highlight the Part in the browser.

Right click and select **Rename.**

93.

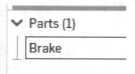

Rename the part to **Brake**.

94. 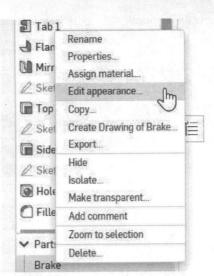 Right click on the Brake in the browser.
Select **Edit Appearance.**

95. 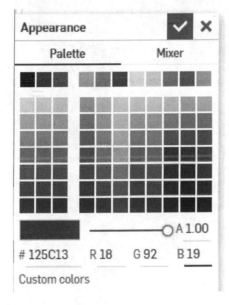 Assign an R value of **18.**

Assign a G value of **92.**

Assign a B value of **19.**

Click the Green check.

96. 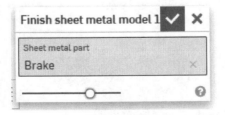 Click **Finish Sheet Metal Part**.

97.  Click the Part Name in the browser.

Green check.

98.

On the right of the display window, locate the tab for **Sheet Metal table and Flat View**.

You may need to perform a refresh on the browser to see the tab.

Click to open the tab.

99.

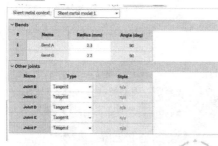

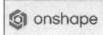
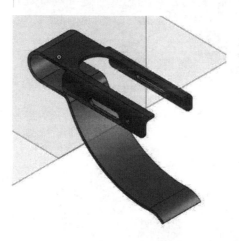

You see the flat pattern as well as the list of bends.

To close the panel, click the tab.

100.

Close the document by clicking on the Onshape logo.

Part Three: The Wheel

Estimated Time: 90 minutes
Objectives:

- Sweep
- Loft
- Revolve
- Circular Pattern
- Boolean Union
- Delete Face
- Assign Appearance
- Rename Part

1.

   Log in to your Onshape account.
   Left click on the **Scooter** document.

2.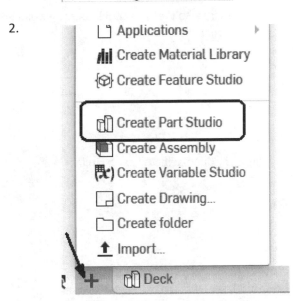

   Select the +.
   Select **Create Part Studio**.

3.

Select **Rename** on the Part Studio tab.

4.

Type **Wheel**.
Click **ENTER**.

5.

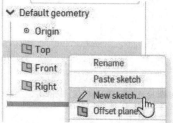

Select the **Top** plane for a new sketch.

6.

Select the **Circle** tool.

7.

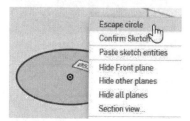

Place a circle so that the center point is coincident to the origin.

8.

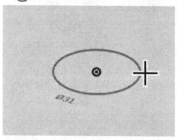

Right click and select **Escape circle**.

9.

Select the **Dimension** tool.

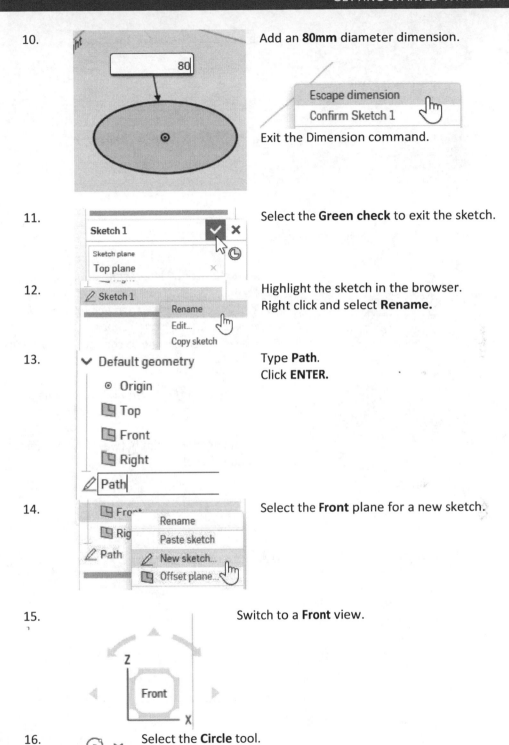

10. Add an **80mm** diameter dimension.

Exit the Dimension command.

11. Select the **Green check** to exit the sketch.

12. Highlight the sketch in the browser.
Right click and select **Rename.**

13. Type **Path**.
Click **ENTER.**

14. Select the **Front** plane for a new sketch.

15. Switch to a **Front** view.

16. Select the **Circle** tool.

17. 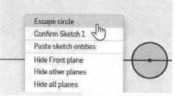 Place a circle to the left of the origin.

18. 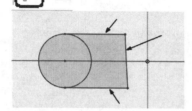 Right click and select **Escape circle**.

19. Select the **Line** tool.

20. 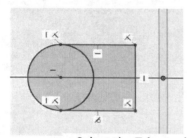 Place two horizontal lines at the upper and lower quadrants.
You can use object tracking to line up the endpoint of the lines with the circle's center.
Draw a short vertical line to connect the two lines.

21. 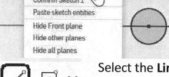 Add a vertical and/or horizontal constraints as needed.

Add a tangent constraint between the horizontal lines and the circle.

22. Select the **Trim** tool.

23. 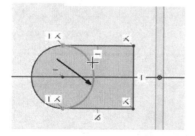 Delete the inside arc.

24. 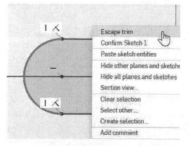 Right click and select **Escape trim**.

25. 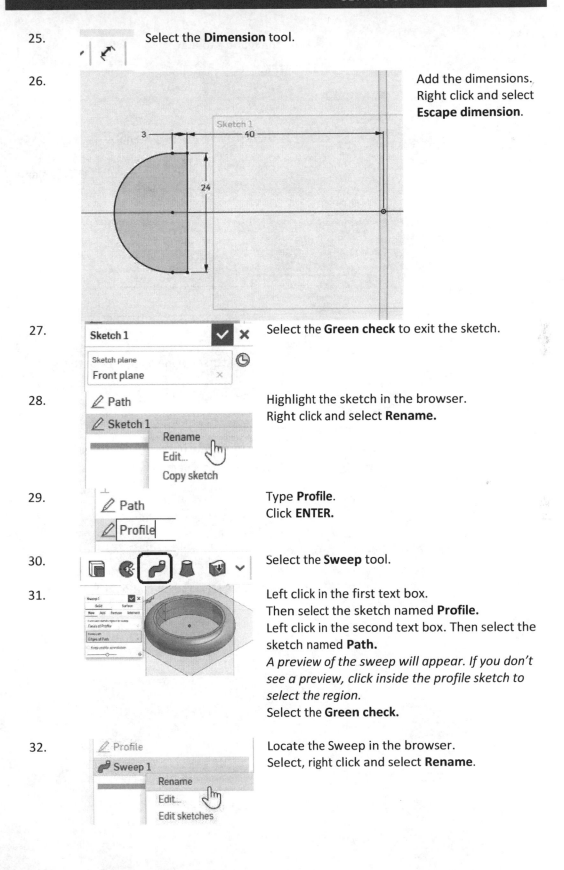 Select the **Dimension** tool.

26. Add the dimensions.
Right click and select
Escape dimension.

27. Select the **Green check** to exit the sketch.

28. Highlight the sketch in the browser.
Right click and select **Rename.**

29. Type **Profile**.
Click **ENTER.**

30. Select the **Sweep** tool.

31. Left click in the first text box.
Then select the sketch named **Profile.**
Left click in the second text box. Then select the
sketch named **Path.**
*A preview of the sweep will appear. If you don't
see a preview, click inside the profile sketch to
select the region.*
Select the **Green check.**

32. Locate the Sweep in the browser.
Select, right click and select **Rename.**

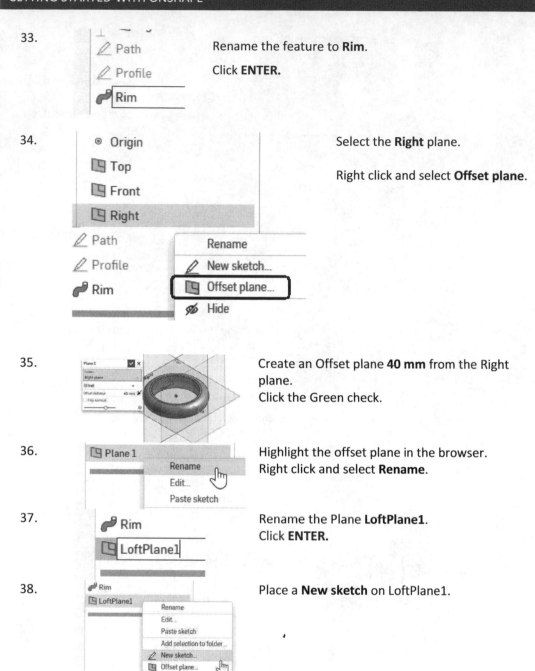

33.

Rename the feature to **Rim**.

Click **ENTER.**

34.

Select the **Right** plane.

Right click and select **Offset plane**.

35.

Create an Offset plane **40 mm** from the Right plane.
Click the Green check.

36.

Highlight the offset plane in the browser.
Right click and select **Rename**.

37.

Rename the Plane **LoftPlane1**.
Click **ENTER.**

38.

Place a **New sketch** on LoftPlane1.

39.

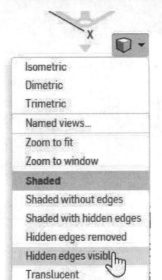

Left click on the Display Options cube.
Select **Hidden edges visible**.

40.

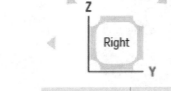

Switch to a **Right** view.

41.

Draw a horizontal line with the midpoint
coincident to the origin.

42.

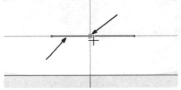

To add a midpoint constraint, select the line,
select the Origin and then select the midpoint
constraint.

43.

Select the **Dimension** tool.

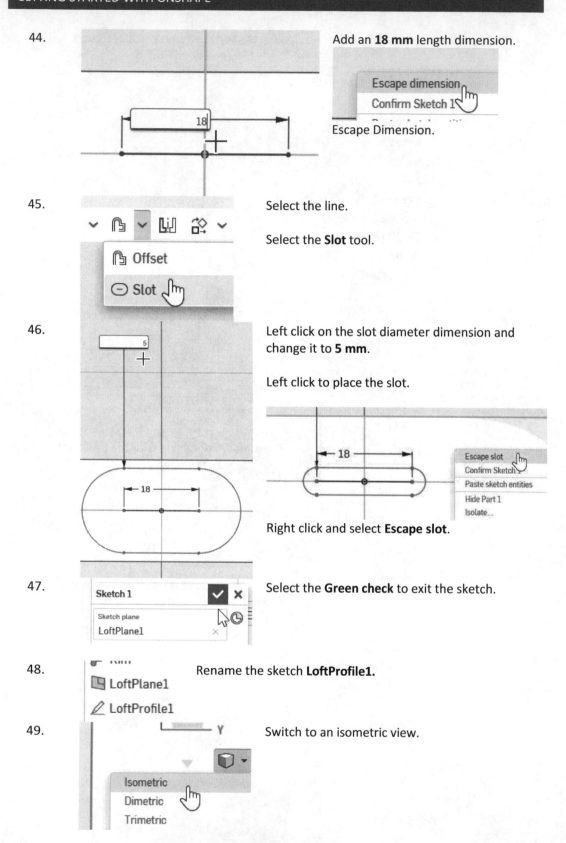

44.   Add an **18 mm** length dimension.

Escape Dimension.

45.   Select the line.

Select the **Slot** tool.

46.   Left click on the slot diameter dimension and change it to **5 mm**.

Left click to place the slot.

Right click and select **Escape slot**.

47.   Select the **Green check** to exit the sketch.

48.   Rename the sketch **LoftProfile1.**

49.   Switch to an isometric view.

50.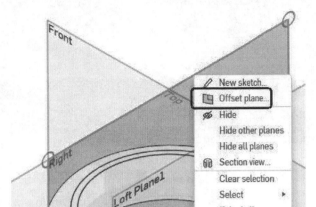

Select the **Right** plane.

Right click and select **Offset plane**.

51.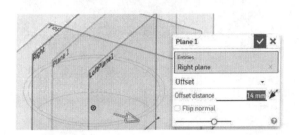

Create an Offset plane **14 mm** from the Right plane.
Click the Green check.

52.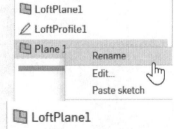

Highlight the offset plane in the browser.
Right click and select **Rename**.

53.

Rename the Plane **LoftPlane2**.
Click ENTER.

54.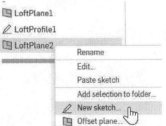

Select LoftPlane2 for a **New sketch.**

55. 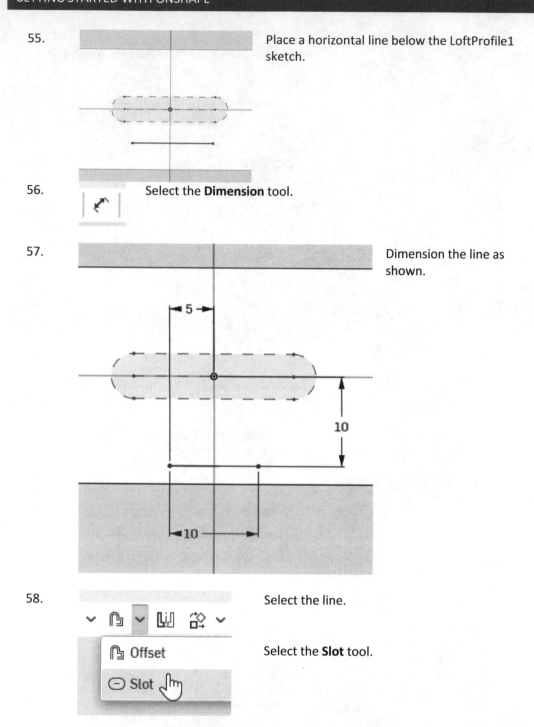 Place a horizontal line below the LoftProfile1 sketch.

56. Select the **Dimension** tool.

57. Dimension the line as shown.

58. Select the line.

Select the **Slot** tool.

Offset

Slot

59.

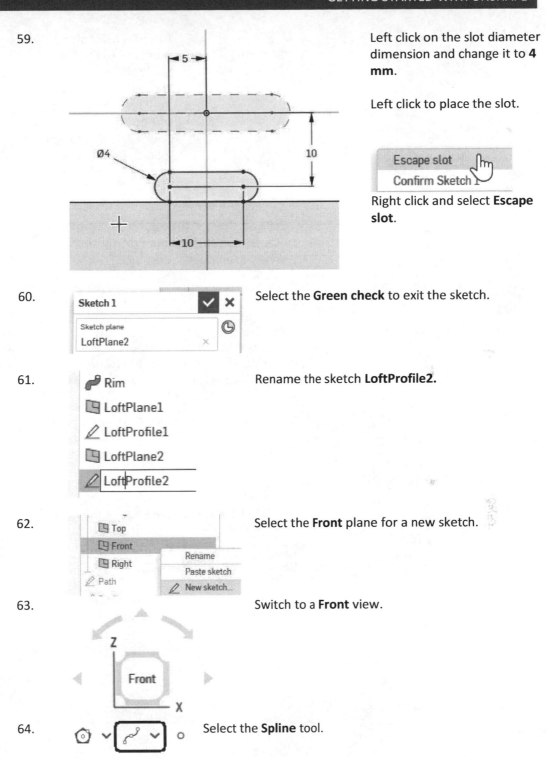

Left click on the slot diameter dimension and change it to **4 mm**.

Left click to place the slot.

Right click and select **Escape slot**.

60.

Select the **Green check** to exit the sketch.

61.

Rename the sketch **LoftProfile2.**

62.

Select the **Front** plane for a new sketch.

63.

Switch to a **Front** view.

64.

Select the **Spline** tool.

65.

The start point of the spline should be to the right of LoftProfile2.
The endpoint of the spline should be to the left of LoftProfile1.
Double left click to end the spline.

66.

Select the bottom point of the LoftProfile2 sketch.
Select the start point of the spline.
Add a coincident constraint.

67.

Select the bottom point of the LoftProfile1 sketch.
Select the end point of the spline.
Add a coincident constraint.

68.

Adjust the spline using the control points to smooth it out.

69.

Select the **Green check** to exit the sketch.

70.

Highlight the sketch in the browser.
Right click and select **Rename.**

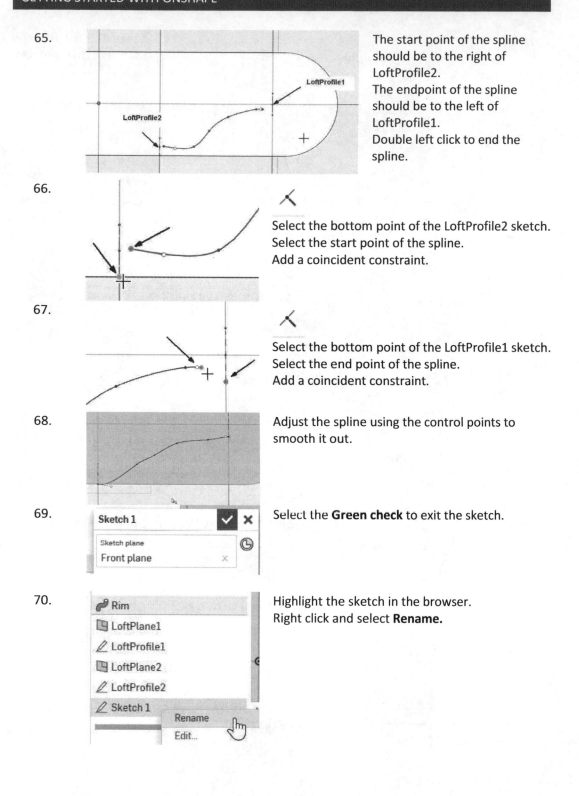

71.

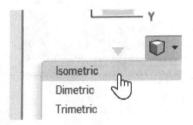

Type **Guide Curve**.

Click ENTER.

There should be three sketches for the loft:
- Guide Curve
- LoftProfile1
- LoftProfile2

Note that the icon for a sketch is a pencil.

Check in the browser to verify you have all three sketches.

72.

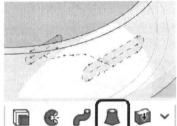

Switch to an isometric view.

You should see the three sketches.

73.

Select the **Loft** tool.

74.

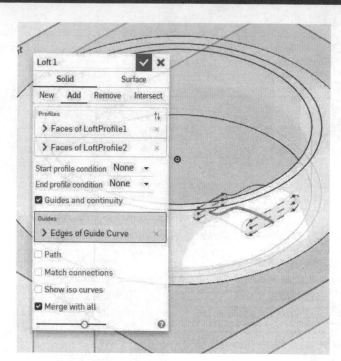

Select **Add**.
Select the two profile sketches.
Enable **Guides and continuity**.
Select the **Guide Curve**.
Enable **Merge with all**.
Click the Green check.

The first spoke is created.
I turned off the visibility of all the planes so I could focus on my part.

To hide planes, select in the browser, right click and select **Hide all planes**.

75.

 Spoke

Rename the Loft **Spoke.**

76.

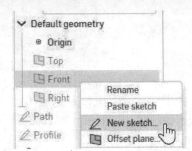

Select the Front plane in the browser.
Right click and select **New sketch**.

77.

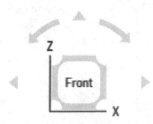

Switch to a **Front** view.

78.

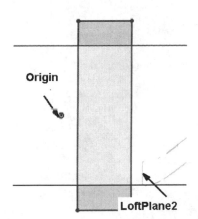

□ ⌄

Draw a corner rectangle between the origin and
LoftPlane2.

79.

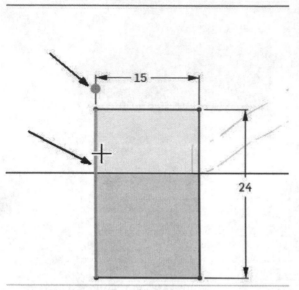

Add a midpoint constraint between the left vertical line of the rectangle and the origin.
Add a 15 mm horizontal dimension.
Add a 24 mm vertical dimension.

80. Select the **Revolve** tool.

81.

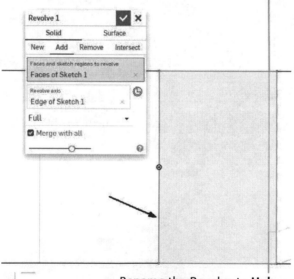

Select **Add**.

Select the left vertical line of the rectangle to use as the axis for the revolve.

Enable **Merge with all.**

Click the Green check.

82.

Rename the Revolve to **Hub.**

Spoke

Sketch 1

Hub

Rim

LoftPlane1

LoftProfile1

LoftPlane2

LoftProfile2

Guide Curve

Spoke

Sketch 1

Hub

Note there are now three features listed in the browser, the rim, the spoke, and the hub.

83.

Inspect the part.

84.

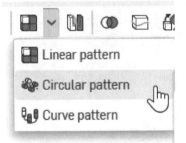

Select **Circular pattern** from the Features ribbon.

85.

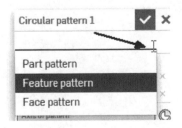

Select **Feature pattern** from the drop-down list.

86.

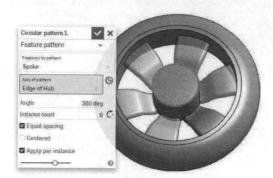

Select the Spoke in the display window or the browser.

Select the circumference edge of the hub as the axis of rotation.

Set the value of the pattern to **360 deg**.

Set the Count to **6**.

Enable **Equal Spacing**.

Enable **Apply per instance**.

Click the Green check.

87.

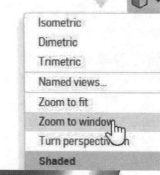

Left click on the Display Options cube.

Select **Zoom to window.**

88.

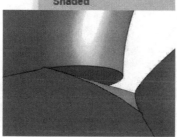

Zoom into the area where the spoke meets the hub.

You see a small face which should be deleted.

89.

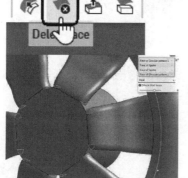

Select **Delete Face** from the ribbon.

90.

Orbit around the model and select all the faces to be deleted.

91.

Delete face 1 ✓ ✕

Face of Circular pattern 1 ✕
Face of Spoke ✕
Face of Spoke ✕
Face of Circular pattern 1 ✕

Heal ▼

☑ Delete fillet faces

Enable **Delete Fillet faces.**
Click the Green check.

92.

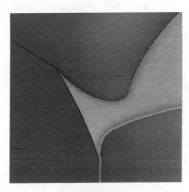

Zoom in and you will see that the spokes now meet the hub cleanly.

93.

Select the top of the hub for a new sketch.

94.

Place a point coincident to the origin.

95.

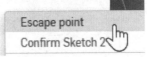

Escape point
Confirm Sketch 2

Right click and select **Escape point**.

96.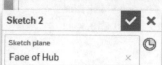

Click the Green check to exit the sketch.

97.  Select the **Hole** tool.

98.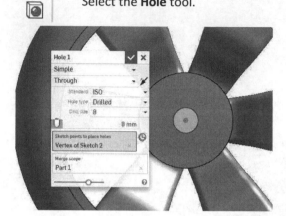

Set the Hole type to **Simple**.

Set the Standard to **ISO**.

Set the Hole diameter to **8 mm**.

Set the Hole termination to **Through**.

Select the point that was just placed.

Click the Green check.

99.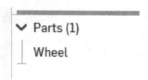

Rename the Part in the browser **Wheel.**

100.

Select the Wheel in the browser.
Right click and select **Edit Appearance**.

101.

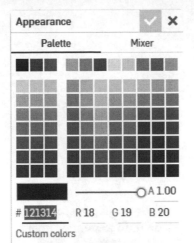

Set the R value to **18**.
Set the G value to **19**.
Set the B value to **20**.
Click the Green check.

102.

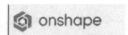

Close the document.

Part Four: The Clamp

Estimated Time: 40 minutes
Objectives:
- Symmetric Extrude
- Counterbore Hole
- Extrude - Remove
- Circular Pattern - Sketch

1.  Log in to your Onshape account.
Left click on the **Scooter** document.

2.   Select the +.
Select **Create Part Studio**.

3. Select **Rename** on the Part Studio tab.

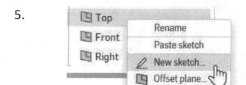

4.  Type **Clamp**.
Click **ENTER**.

5. Select the **Top** plane for a new sketch.

6. Use the Orientation cube to switch to a **TOP** view.

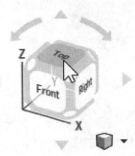

7. 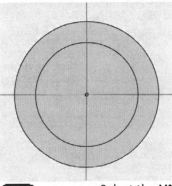 Draw two concentric circles with the center points coincident to the origin.

8. Select the **LINE** tool.

9. 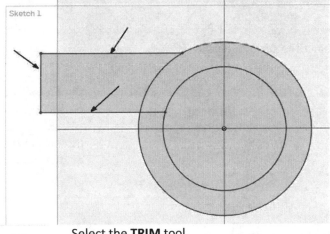 Draw three lines with the inside end point of the bottom line coincident to the inside circle and the inside end point of the top line coincident to the outside circle.

10. Select the **TRIM** tool.

11.

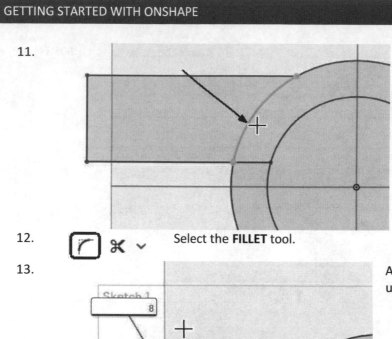

Use the Trim tool to remove the inside arc indicated.

12.

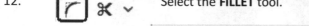

Select the **FILLET** tool.

13.

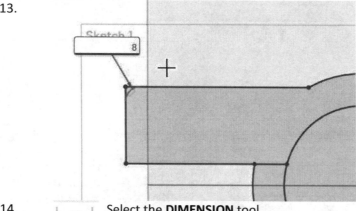

Add an **8 mm** fillet to the upper corner.

14.

Select the **DIMENSION** tool.

15.

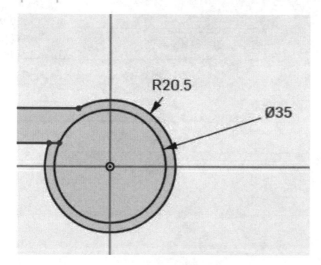

Dimension the outer circle radius **20.5 mm**.

Dimension the inner circle diameter **35 mm**.

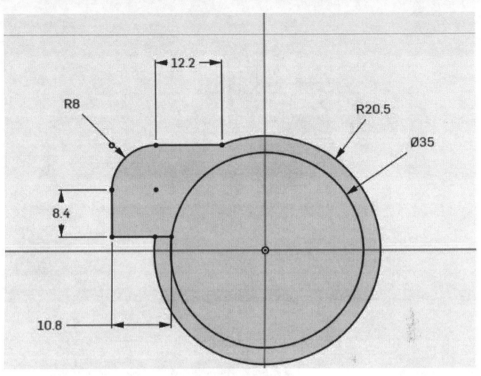

16. Dimension the top horizontal line **12.2 mm.**

Dimension the bottom horizontal line **10.8 mm**.

Dimension the vertical line **8.4 mm**.

This should fully define the sketch.

17. Enable **Construction** mode.

18. Select the **LINE** tool.

19. Place a horizontal construction line starting at the origin and ending to the right.

20.

12.2
R8
R20.5
Ø35
8.4
10.8

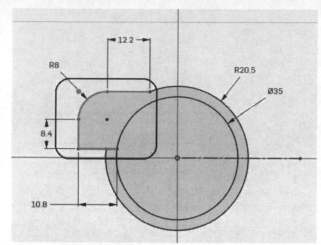

Select the fillet and three lines.

21. Select the **Mirror** tool.

22.

12.2
R8
R20.5
Ø35
8.4

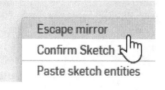

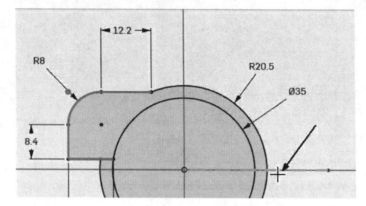

Select the horizontal construction line.

23.

Escape mirror
Confirm Sketch 1
Paste sketch entities

Right click and select **Escape Mirror**.

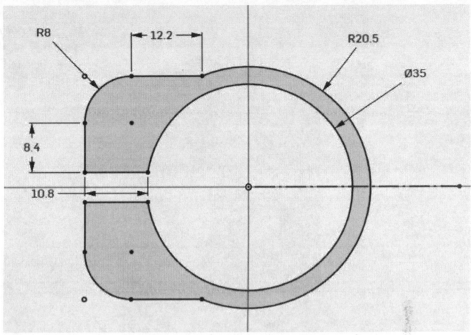

24.   Use the **Trim** tool to clean up the sketch and remove the inner lines.

25. 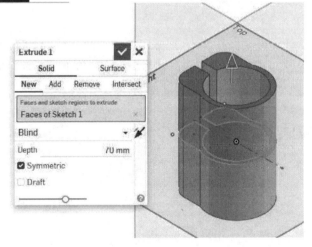 Select the **Extrude** tool.

26.   Select the **New** option.
Enable **Symmetric**.
Set the Depth to **70 mm**.
Select the **Green check**.

27.

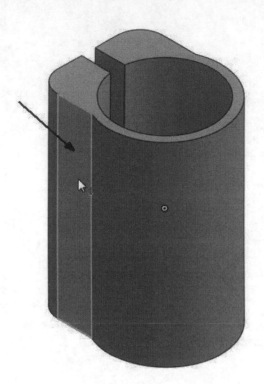

Select the face indicated. *This is the flat face that measures 12.2 mm.* Right click and select **New sketch.**

28.

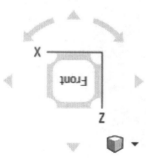

Use your Orientation cube to adjust the view so you are looking at the Front face upside down.

Tip: Use the Rotation arrows.

29.

Select the **POINT** tool.

30.

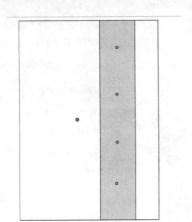

Place four points.

31.

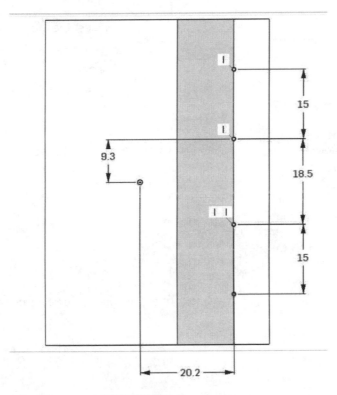

Align them vertically using a vertical constraint.

Dimension as shown.

32.

Exit the sketch by selecting the green check.

33.   Select the **Hole** tool.

34.

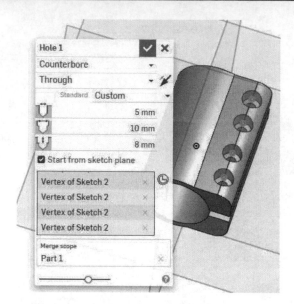

Set the Hole type to **Counterbore**.

Set the Standard to **Custom**.

Set the hole diameter to **5 mm**.

Set the Counterbore diameter to **10 mm**.

Set the Counterbore depth to **8 mm**.

Enable **Start from sketch plane**.

Set the hole termination to **Through**.

Select the four points.

Select **Part 1** for the Merge scope.

Click the Green check.

35.

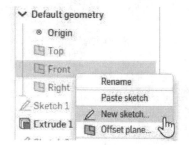

Select the **Front** plane in the browser.
Right click and select **New sketch.**

36.

Orient to a Front view – correct side up.

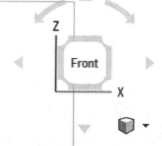

37.

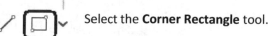

Select the **Corner Rectangle** tool.

38.

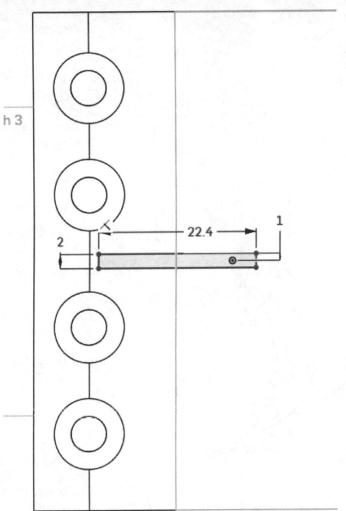

Draw a corner rectangle that is **22.4 mm** long and **2 mm** wide.
Center it in the width direction to the origin.

39.

Select the edge of the part and the short edge of the rectangle.

Select **Coincident** constraint.

The sketch is now fully defined.

2

40. Select the **Extrude** tool.

41.

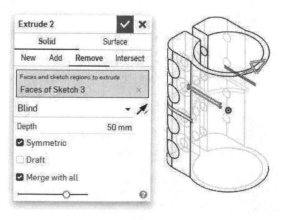

Extrude 2

Solid | Surface

New | Add | **Remove** | Intersect

Faces and sketch regions to extrude
Faces of Sketch 3

Blind

Depth | 50 mm

☑ Symmetric

☐ Draft

☑ Merge with all

Enable **Remove**.

Enable **Symmetric**.

Set the Distance to **50 mm**.

Enable **Merge with all.**

Click the Green check.

42.

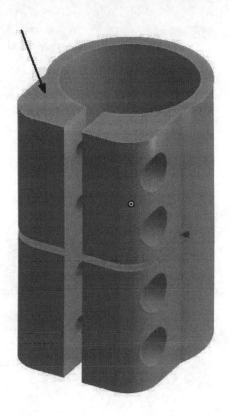

Select the top face for a new sketch.

43.

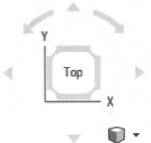

Switch to a **TOP** view.

44. 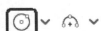   Select the **CIRCLE** tool.

45.

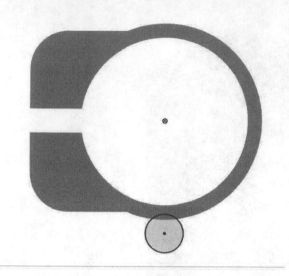

Draw a centerpoint circle below the origin and slightly outside the part.

Align the circle's centerpoint vertically to the origin using a vertical constraint.

46.

Select the **DIMENSION** tool.

47.

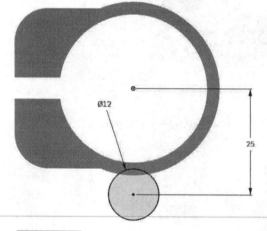

Add a vertical distance of **25 mm**.
Add a diameter of **12 mm**.

48.

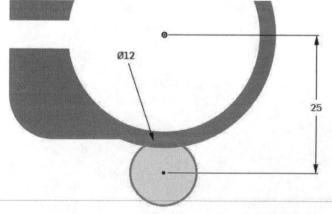

Select the circle.

49.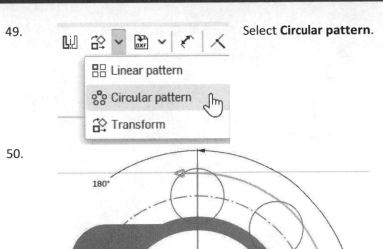
Select **Circular pattern**.

Linear pattern

Circular pattern

Transform

50. Set the count to **5x.**
Set the angle to fill to **180°.**
Left click on the arrow to change the direction of the pattern.

Click the left mouse button to accept.

180°

Ø12

25

5x

51. Select the **Extrude** tool.

52. Enable **Remove**.
Select **Blind**.
Set the Distance to **8 mm**.
Enable **Merge with all**.
Click the Green check.

Extrude 3

Solid | Surface

New | Add | Remove | Intersect

Faces and sketch regions to extrude
Faces of Sketch 4

Blind

Depth | 8 mm

Symmetric

Draft

Second end position

Merge with all

53.

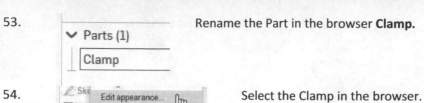

Rename the Part in the browser **Clamp.**

54.

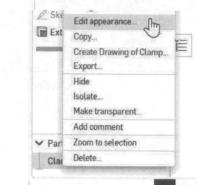

Select the Clamp in the browser.
Right click and select **Edit Appearance**.

55.

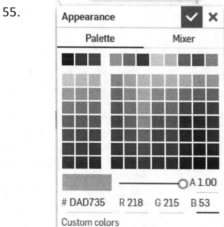

Set the R value to **218**.
Set the G value to **215**.
Set the B value to **53**.
Click the Green check.

56.

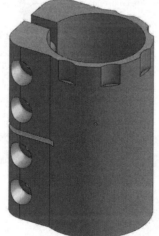

Close the document.

Part Five: The Fork

Estimated Time: 60 minutes

Objectives:

- Extrude
- Simple Hole
- Ellipse
- Use (Project/Convert)
- Extrude - Remove
- Mirror
- Chamfer

1.  Log in to your Onshape account.
   Left click on the **Scooter** document.

2. 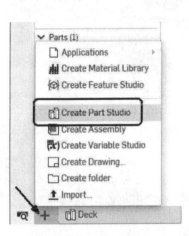 Select the +.
   Select **Create Part Studio**.

3.  Select **Rename** on the Part Studio tab.

4. 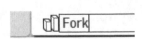 Type **Fork**.
   Click **ENTER.**

5. Select the **Top** plane for a new sketch.

6. Select the **Circle** tool from the ribbon.

7. Draw a circle coincident to the origin.

8. Right click and select **Escape circle**.

9. Select the **Dimension** tool.

10. Add a **28 mm** diameter dimension.

    If the circle doesn't display as black/fully constrained, add a coincident constraint between the origin and the circle's center point.

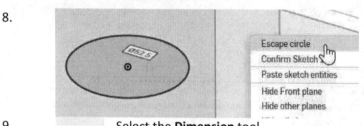

11. Select the **Extrude** tool.

12. Select **New**.
    Set the Termination to **Blind**.
    Set the Distance to **144 mm**.
    Click the Green check.

13. Select the **Extrude i**n the browser.
Right click and select **Rename**.

14. Rename the feature to **Shaft.**

Click **ENTER.**

15. Select the **Top** plane for a new sketch.

16. Select the **Center point rectangle** tool.

17. 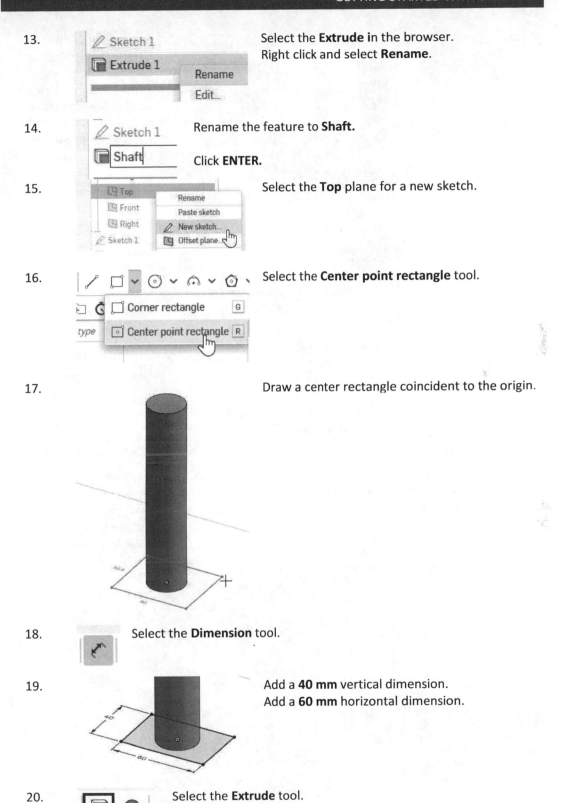 Draw a center rectangle coincident to the origin.

18. Select the **Dimension** tool.

19. Add a **40 mm** vertical dimension.
Add a **60 mm** horizontal dimension.

20. Select the **Extrude** tool.

21.

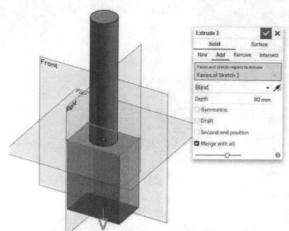

Select **Add**.
Set the Termination to **Blind**.
Set the Distance to **80 mm**.
Change the direction to down.
Enable **Merge with all**.
Click the Green check.

22.

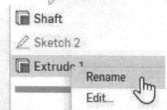

Select the **Extrude i**n the browser.
Right click and select **Rename**.

23.

Shaft
Sketch 2
Fork

Rename the part to **Fork**.
Click **ENTER.**

24.

Select the **Fillet** tool.

25.

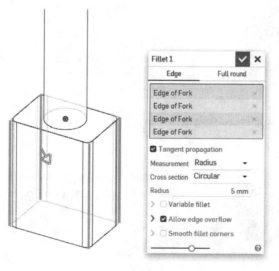

Add a **5 mm** fillet to the vertical edges of the fork.

I changed the display to Hidden Edges visible, so you can see the four vertical edges to select.

26.

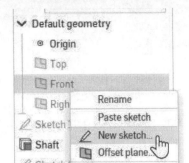

Select the **Front** plane for a new sketch.

27.

Switch to a **Front** view.

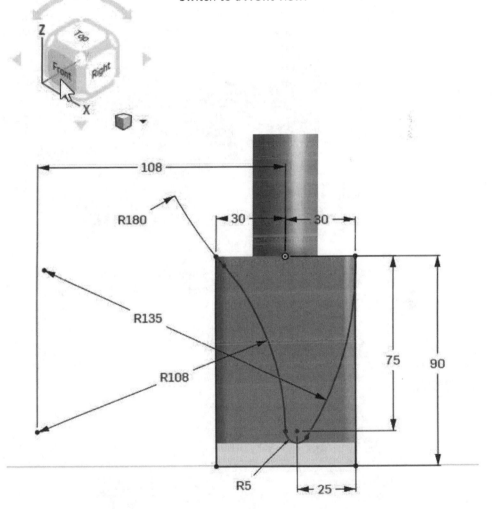

28.

Create the sketch.

The sketch has four arcs and three lines (two vertical lines and one horizontal line).

There is a tangent constraint between the bottom 10 mm diameter arc and the two connecting arcs.

There is a tangent constraint between the two left side arcs.

29.

Select the **Extrude** tool.

30.

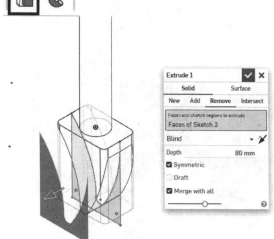

Select **Remove**.
Set the Termination to **Symmetric**.
Set the Distance to **80mm**.

Enable **Merge with all**.

Click the Green check.

31.

Select the Extrude in the browser.
Right click and select **Rename**.

32.

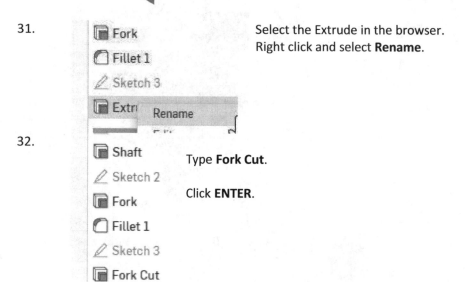

Type **Fork Cut**.

Click **ENTER**.

33. 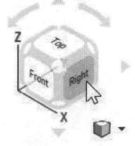 Select the **Right** plane for a new sketch.

34. 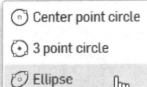 Change to a **Right** view.

35. Select the **Ellipse** tool.

36.

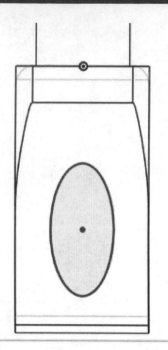

Place an ellipse with the long axis on the vertical side.

37.

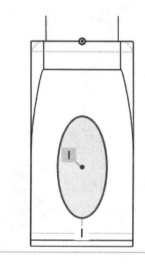

Add a vertical constraint between the center of the ellipse and the origin.

38.   Select the **LINE** tool.

39.

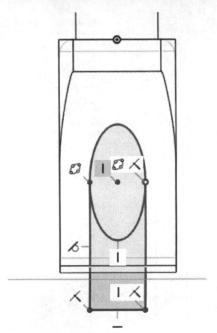

Add a vertical line to each side of the ellipse and a horizontal line at the bottom of the vertical lines.

40. ✂ ⌄   Select the **Trim** tool.

41.

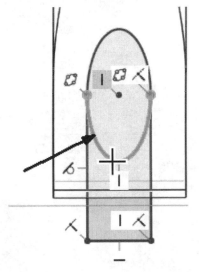

Trim out the bottom of the ellipse.

42. ↰   Select the **Dimension** tool.

43.

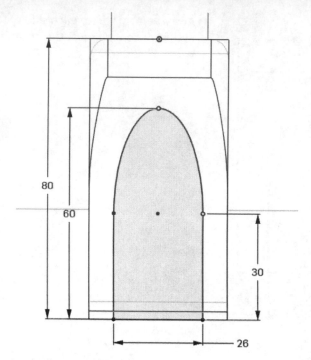

Add the dimensions shown.

I placed a point at the top vertex of the ellipse in order to create the 60 mm vertical dimension.

80

60

30

26

44. Select the **Extrude** tool.

45.

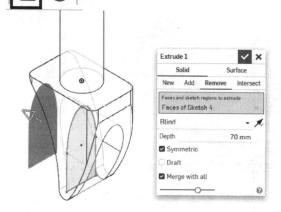

Select **Remove**.
Set the Termination to **Symmetric**.
Set the Distance to **70mm**.
Enable **Merge with all**.

Click the Green check.

46.

Shaft

Sketch 2

Fork

Fillet 1

Sketch 3

Fork Cut

Sketch 4

Opening

Rename the Extrude **Opening**.

47.

Use the Orientation Cube to change the view display to **Right-Back** isometric.

48.

Select the Back face of the fork for a new sketch.

49.

Switch to a **Back** view.

50.

Select the **Point** tool.

51.

Place a point.
Make it concentric to the bottom arc.

Sketch 5   ✓ ✗

Sketch plane
Face of Fork   ✕

Exit the sketch by selecting the Green check.

52.

Select the **Hole** tool.

53.

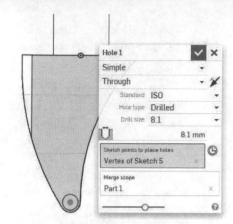

Set the Hole type to **Simple**.
Set the hole diameter to **8.1 mm**.
Set the hole termination to **Through**.
Select the point.
Select **Part 1** for the Merge scope.
Click the Green check.

54.

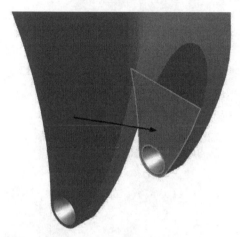

Select the inside face of the fork for a new sketch.

55.

Select the outer edge of the hole.

56.

Select **Use (Project/Convert)**.

57.

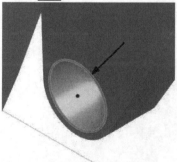

Select the copied edge.

58.  Select **Offset**.

59. 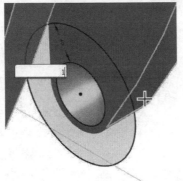 Change the offset dimension to **1 mm**.

60. Select the **Extrude** tool.

61. 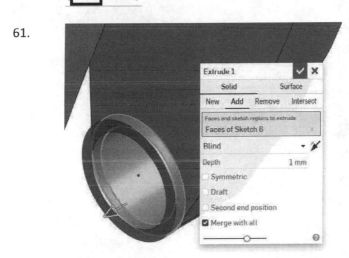 Select **Add.**
Set the Termination to **Blind**.
Set the Distance to **1 mm**.
Enable **Merge with all**.
Click the Green check.

62.  Rename the Extrude feature **Boss.**

63. Select the **Mirror** tool.

64.

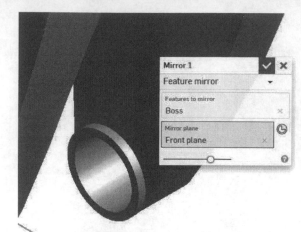

Select **Feature Mirror** from the drop-down list.
Select **Boss** as the feature to be mirrored.
Select the **Front plane**.
You should see a preview of the new mirrored feature.
Click the Green check.

65.

Select the top of the shaft for a new sketch.

66.

Select the **Point** tool.

67.

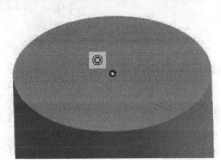

Place a point concentric to the cylindrical face.

68.

Exit the sketch by selecting the Green check.

69. Select the **Hole** tool.

70.

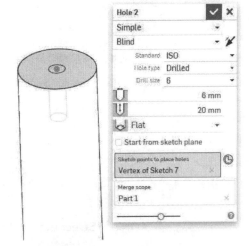

Set the Hole type to **Simple**.
Set the hole diameter to **6 mm**.
Set the hole termination to **Blind.**
Set the Distance to **20 mm.**
Set the hole termination to **Flat.**
Select the point.
Select **Part 1** for the Merge scope.
Click the Green check.

I changed the display to Hidden Edges Visible so you can preview the hole.

71.

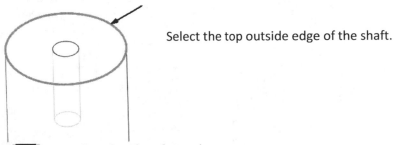

Select the top outside edge of the shaft.

72. Select the **Chamfer** tool.

73.

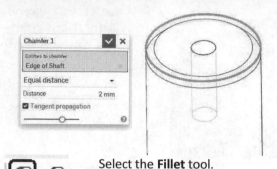

Set the Chamfer type to **Equal Distance**.
Set the Distance to **2 mm**.
Click the Green check.

74.

Select the **Fillet** tool.

75.

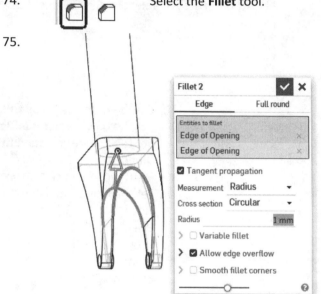

Select the inside edges of the ellipse in the fork opening.
Set the fillet radius to **1 mm.**
Click the Green check.

76.

Select the **Fillet** tool.

77.

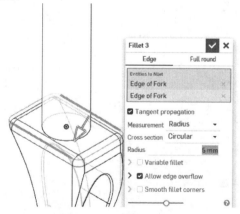

Select the front and back edge of the fork.
Set the fillet radius to **5 mm.**
Click the Green check.

78.

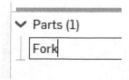

Rename the Part in the browser **Fork.**

79.

Select the Fork in the browser.
Right click and select **Edit Appearance**.

80.

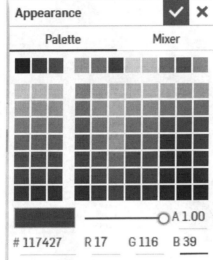

Set the R value to **17.**
Set the G value to **116**.
Set the B value to **39.**
Click the Green check.

81.

Close the document.

Part Six: Handlebars

Estimated Time: 50 minutes
Objectives:
- Extrude
- Shell
- Reference Plane using Point Plane
- Sweep
- Boolean
- Revolve

1. Log in to your Onshape account.
    Left click on the **Scooter** document.

2.  Select the +.
    Select **Create Part Studio**.

3. Select **Rename** on the Part Studio tab.

4.  Type **Handlebars**.
    Click **ENTER.**

5. 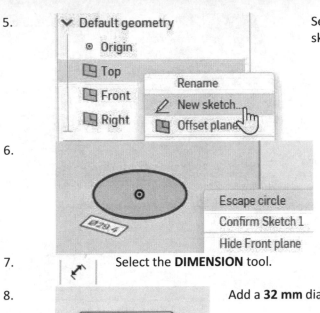 Select the **Top** plane for a new sketch.

6. Draw a **center point circle** coincident to the origin. Right click and select **Escape circle.**

7. Select the **DIMENSION** tool.

8. Add a **32 mm** diameter dimension to the circle.

9. Select the **Extrude** tool.

10.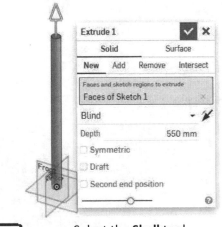

Select **New**.
Set the Termination to **Blind.**
Set the Distance to **550 mm**.
Click the Green check.

11. Select the **Shell** tool.

12.

Select the bottom face to be removed.
Set the wall thickness to **2 mm**.
Click the Green check.
The Hollow option is used if you want to create a hollow part with no faces removed.

13.

To verify that you selected the correct face (BOTTOM) to be removed, check the Orientation cube.

14.

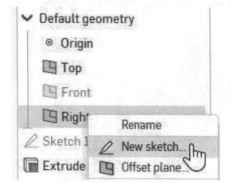

Select the **Right** plane for a new sketch.

15.

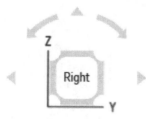

Switch to a **Right** view.

16.

Select **Corner Rectangle**.

17.

Place a corner rectangle where the bottom line is coincident to the origin and the width is centered on the origin.

–●–

Hint: Add a midpoint constraint on the bottom horizontal line and the origin.

18.

Select the **DIMENSION** tool.

19.

Add dimensions to further define the sketch.

Add a 45 mm vertical dimension and a 6 mm width dimension.

20.

Select the **FILLET** tool.

21.

Add a **3 mm** radius fillet to the top of the rectangle.

6

R3   R3

22.

Select the **Extrude** tool.

23.

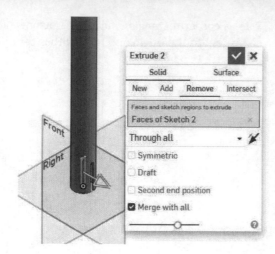

Select **Remove**.
Set the Termination to **Through All**.
Change the direction to remove the material in the positive X direction.
Enable **Merge with all**.
Click the Green check.

Hint: Use the Orientation Cube to help you verify the positive X direction.

24.

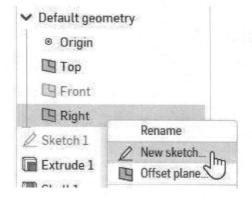

Select the **Right** plane for a new sketch.

25.

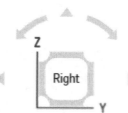

Switch to a **Right** view.

26.

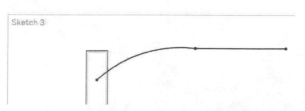

Draw an arc and a horizontal line near the top of the cylinder.

27.

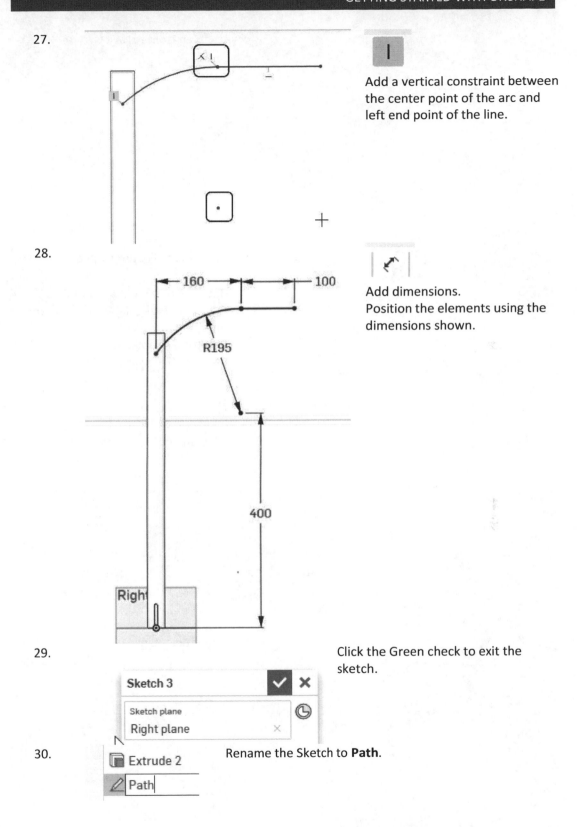

Add a vertical constraint between the center point of the arc and left end point of the line.

28.

Add dimensions.
Position the elements using the dimensions shown.

29.

Click the Green check to exit the sketch.

Sketch 3

Sketch plane
Right plane

30.

Extrude 2

Path

Rename the Sketch to **Path**.

31.

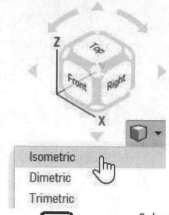

Switch to an **Isometric** view.

32.

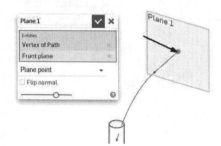

Select the **Plane** tool from the Features ribbon.

33.

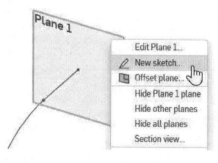

Set the Option to **Plane Point**.
Select the end point of the horizontal line.
Select the **Front** plane.
Click the Green check.

34.

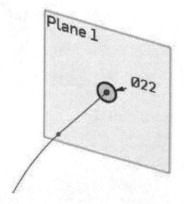

Select the new plane for a new sketch.

35.

 ⌄   Select the **CIRCLE** tool.

36.

Draw a center point circle.

Add a 22 mm diameter.

If the circle is not fully constrained:

Select the end point of the path.
Select the center point of the circle.
Select the **Coincident** tool from the ribbon.
The circle will move to the path.

37.  Click the Green check to exit the sketch.

38. Rename the Sketch **Profile**.

      🔲 Extrude 2

      ✏️ Path

      🔳 Plane 1

      ✏️ Profile

39. Select the **Sweep** tool.

40.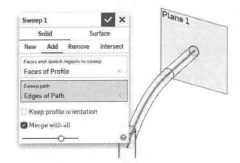

Select the **Add** option.

Select the **Profile** sketch for the profile text box.

Select the **Path** sketch for the path text box.

Enable **Merge with all.**

Click the Green check.

41. Select the **Shell** tool.

42.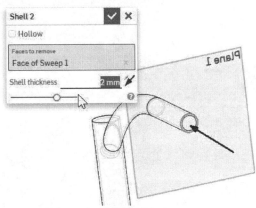

Select the end face of the Sweep to be removed.
Set the Wall thickness to 2 mm.
Click the Green check.

43. Select the **Mirror** tool.

44.

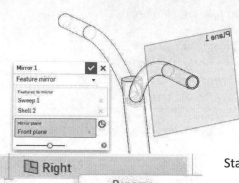

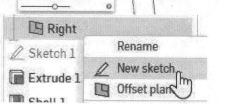

Select **Feature Mirror** from the drop-down list.
Select the Sweep and the Shell from the browser.
Select the **Front** Plane.
Click the Green check.

45.

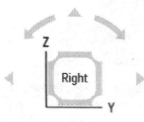

Start a new sketch on the **Right** plane.

46.

Switch to a **Right** view.

47. Enable **Construction** mode.

48.

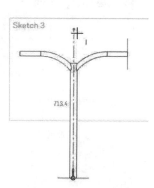

Draw a vertical line starting from the origin and up through the part.

This will be a mirror line.

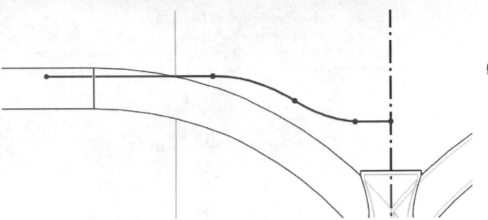

49.   Starting from the left:
       Draw a horizontal line, then a concave arc, then a convex arc, then a horizontal line.

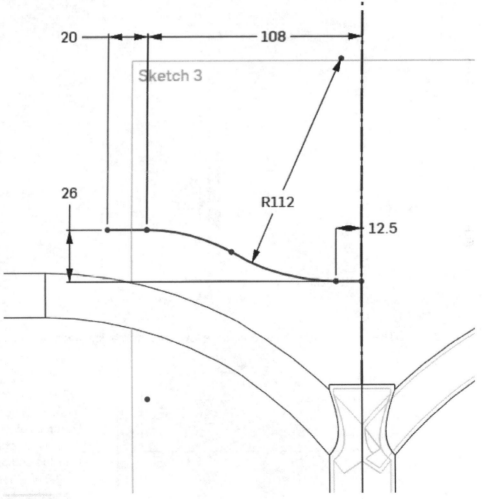

50.

       Add dimensions.
       The radius for the convex arc is 112 mm.

The length of the first horizontal line is 20 mm.
The length of the inner horizontal line is 12.5 mm.
The vertical distance between the two horizontal lines is 26 mm.
The horizontal distance between the inner endpoints of the horizontal lines is 108 mm.

51.

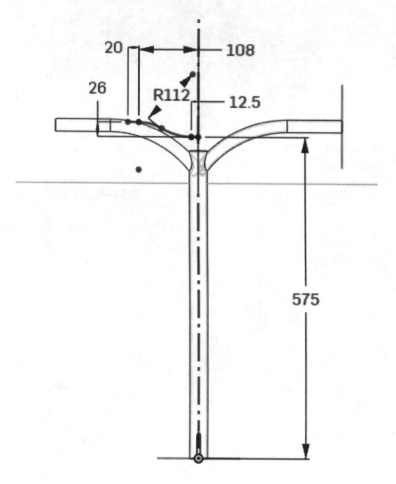

52. The vertical distance between the lower horizontal line and the origin is 575 mm.

53.

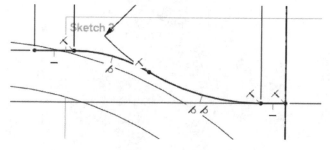

The arcs are tangent to their connecting lines. Add a coincident constraint between the inside end point of the 12.5 mm horizontal line and the vertical construction line, if needed.

54. 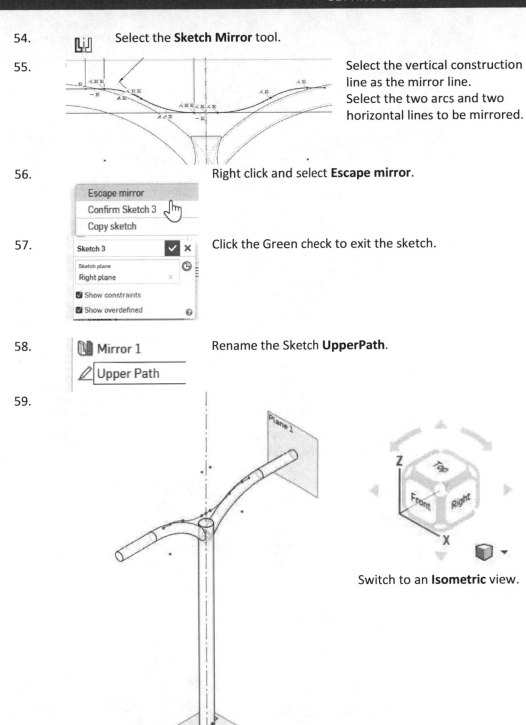 Select the **Sketch Mirror** tool.

55. Select the vertical construction line as the mirror line.
Select the two arcs and two horizontal lines to be mirrored.

56. Right click and select **Escape mirror**.

Escape mirror
Confirm Sketch 3
Copy sketch

57. Click the Green check to exit the sketch.

Sketch 3
Sketch plane
Right plane
☑ Show constraints
☑ Show overdefined

58. Mirror 1
Upper Path

Rename the Sketch **UpperPath**.

59. Switch to an **Isometric** view.

60. Select the **Plane** tool from the Features ribbon.

61.

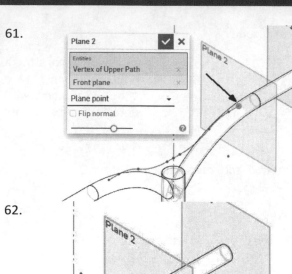

Set the Option to **Plane Point**.
Select the end point of the horizontal line.
Select the **Front** plane.
Click the Green check.

62.

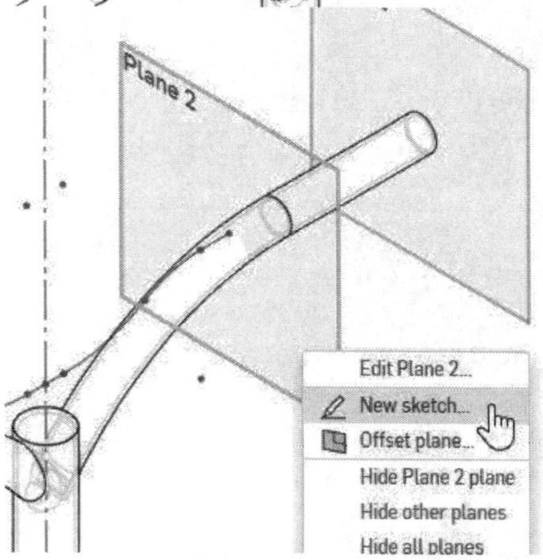

Select the new plane for a new sketch.

63.

Select the **CIRCLE** tool.

64.

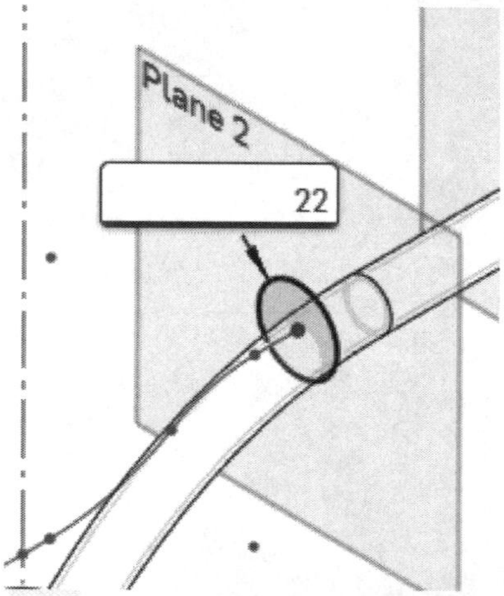

Draw a center point circle.

Add a 22 mm diameter.

If the circle is not fully constrained:

Select the end point of the path.
Select the center point of the circle.
Select the **Coincident** tool from the ribbon.
The circle will move to the path.

65. Click the Green check to exit the sketch.

66. Rename the Sketch to **UpperProfile**.

67. Select the **Sweep** tool.

68.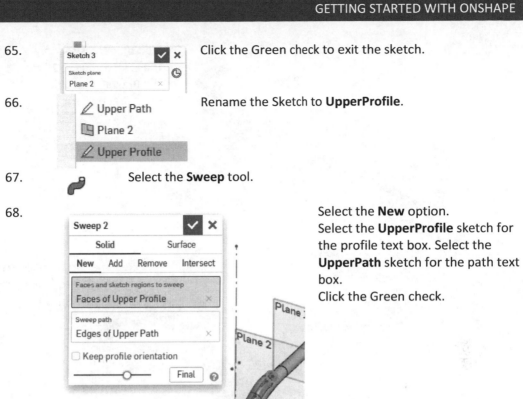

Select the **New** option.
Select the **UpperProfile** sketch for the profile text box. Select the **UpperPath** sketch for the path text box.
Click the Green check.

Notice there are two parts in the browser.
Part 2 is the top bar.
Part 1 is the rest of the model.
If you click on the part name, it will highlight in the window, so you can identify which is which.

This only happens if you selected New and not Add as the option.

69.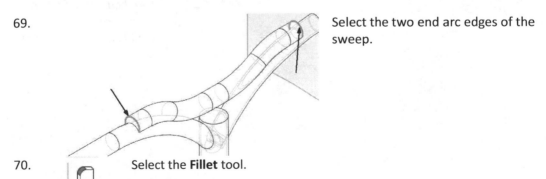

Select the two end arc edges of the sweep.

70. Select the **Fillet** tool.

71.

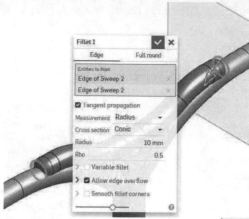

Enable Conic cross section.
Set the Fillet radius to **10 mm**.
Set the Rhovalue to **0.5**.
Click the Green check.

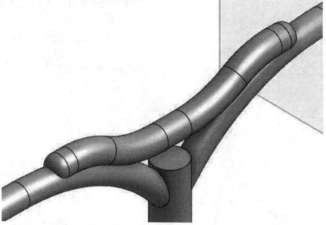

The top bar should be filleted to look more finished.

72. Select the **Shell** tool.

73.

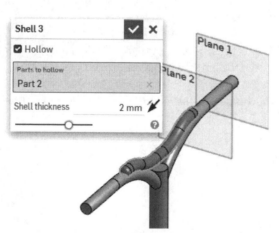

Enable **Hollow**.
Select the top bar (Part 2).
Set the Wall thickness to **2 mm.**
Click the Green check.

If we had not defined the top bar as a separate bar, we would not have been able to hollow it out.

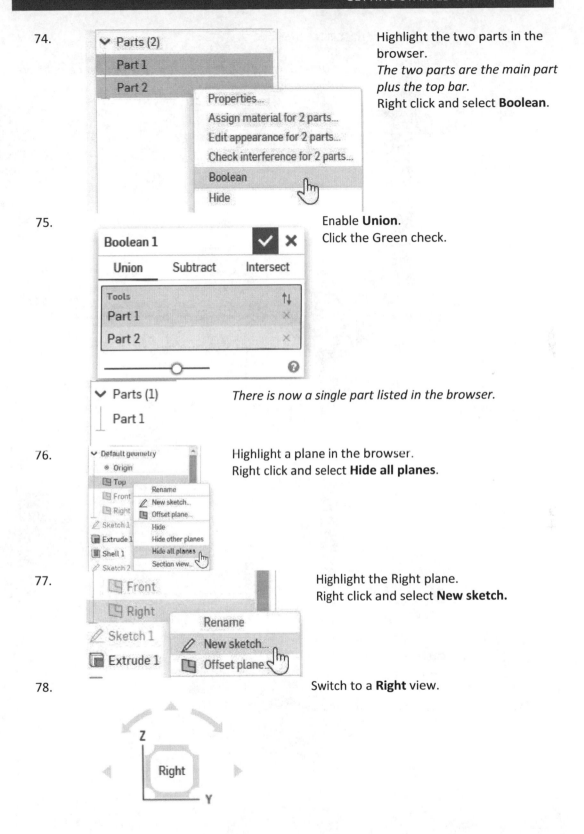

74.

Highlight the two parts in the browser.
The two parts are the main part plus the top bar.
Right click and select **Boolean**.

75.

Enable **Union**.
Click the Green check.

There is now a single part listed in the browser.

76.

Highlight a plane in the browser.
Right click and select **Hide all planes**.

77.

Highlight the Right plane.
Right click and select **New sketch.**

78.

Switch to a **Right** view.

79. Select the **CIRCLE** tool.

80. Draw a center point circle with a **40 mm** diameter above the origin.

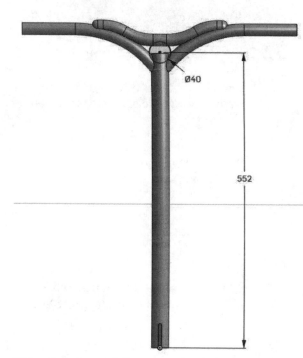

81. Place a vertical dimension of **552 mm** between the circle's center point and the origin.

82. If the circle is not fully constrained:
     Add a vertical constraint between the circle and the origin.

83. Select the **LINE** tool.

84.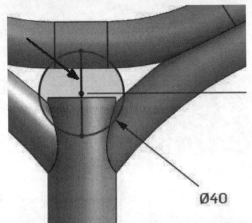

Draw a vertical line bisecting the circle.

Ø40

85.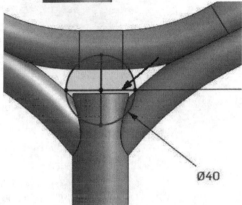

Draw a horizontal line bisecting the circle.

Ø40

86. Select the **TRIM** tool.

87.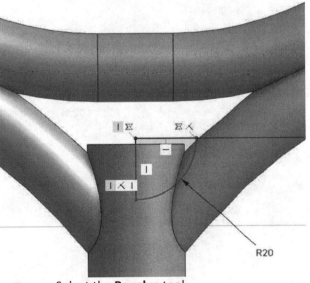

Use **Trim** to create a bottom quarter circle.

R20

88. Select the **Revolve** tool.

89.

Select the **New** option.
Select the vertical line as the axis.
Set the Termination to **Full**.
Click the Green check.

90. Select the **Shell** tool.

91.

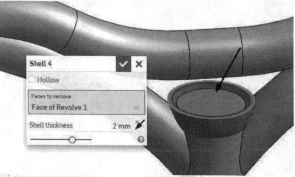

Select the top face of the revolve.
Set the Wall thickness to **2 mm.**
Click the Green check.

92.

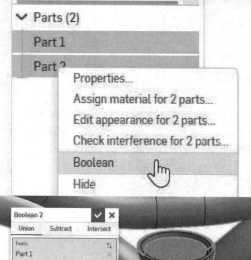

Highlight the two parts in the browser.
The two parts are the main part plus the revolve with shell.
Right click and select **Boolean**.

93.

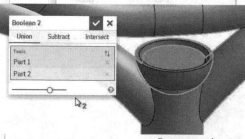

Enable **Union**.
Click the Green check.

94.

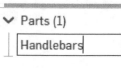

Rename the Part **Handlebars**.

95.

Select the Handlebars in the browser.
Right click and select **Edit Appearance**.

96.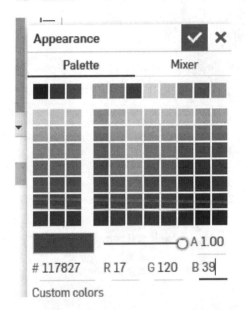

Set the R value to **17.**
Set the G value to **120**.
Set the B value to **39**.
Click the Green check.

97.

Inspect your part and then close
the document.

Part Seven: Hand Grip

Estimated Time: 30 minutes
Objectives:

- Revolve
- Extrude to Face
- Fillet
- Edit Appearance
- Circular Pattern

1.  Log in to your Onshape account.
Left click on the **Scooter** document.

2. 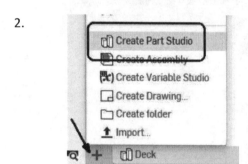 Select the +.
Select **Create Part Studio**.

3. 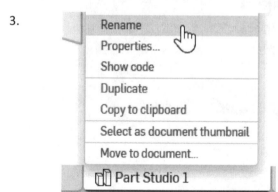 Select **Rename** on the Part Studio tab.

4. Type **Hand Grip**.
Click **ENTER.**

5.

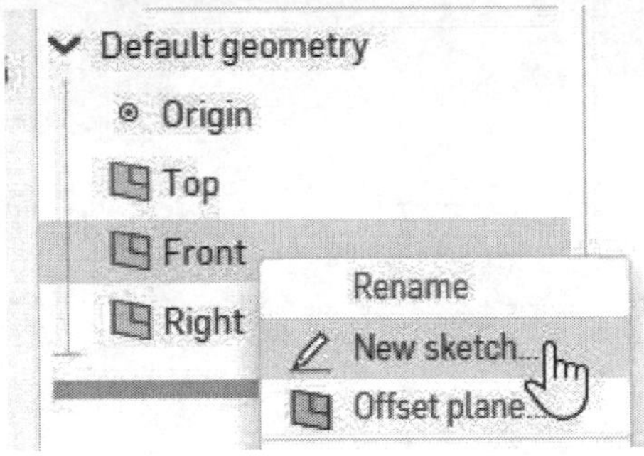

Select the **Front** plane for a new sketch.

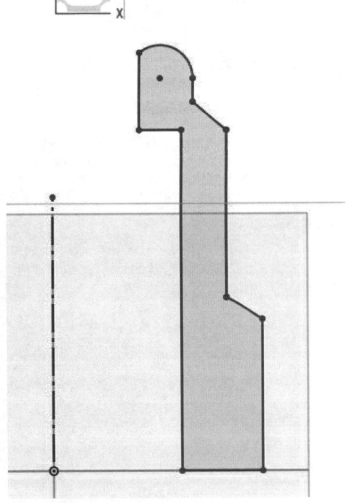

6.

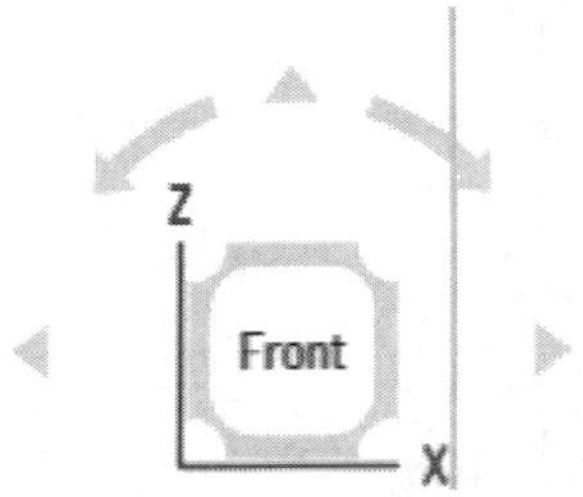

Switch to a **Front** view.

7.

Create the sketch shown.

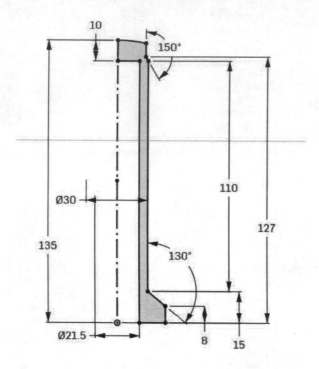

8. Place dimensions as shown.

9. There is a vertical construction line with the end point at the origin.
   There is a small arc at the top. Do not apply a tangent constraint to the arc.
   The center of the arc is coincident to the construction line.
   The left endpoint of the arc is aligned to the origin and is 135 mm above the origin.

10. Select the **Revolve** tool.

11.

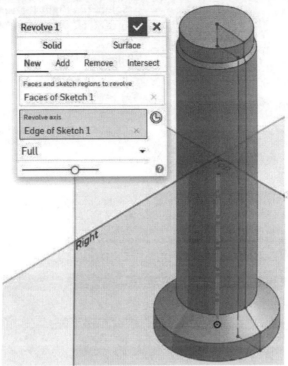

The axis will be the vertical construction line.
Select the **New** option.
Select the vertical construction line.
Select **Full** from the drop-down list.
Click the Green check.

12. Select the **Fillet** tool.

13.

Set the radius to **1.6mm**.
Select all the horizontal circular edges.
Click the Green check.

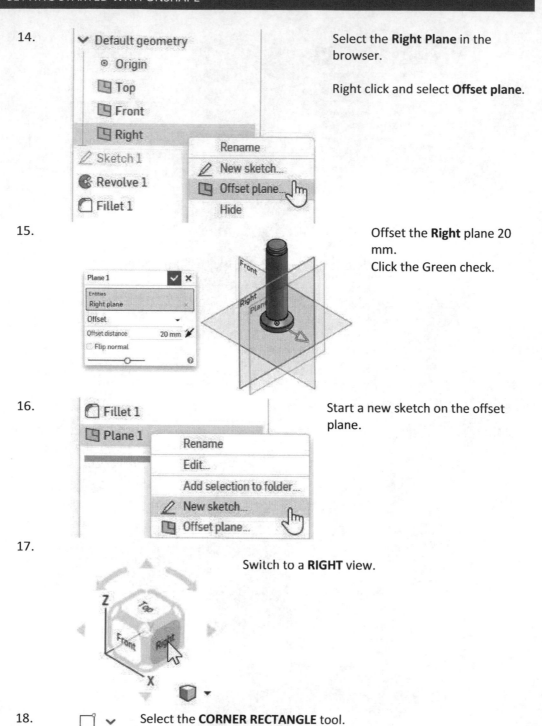

14.

Select the **Right Plane** in the browser.

Right click and select **Offset plane**.

15.

Offset the **Right** plane 20 mm.
Click the Green check.

16.

Start a new sketch on the offset plane.

17.

Switch to a **RIGHT** view.

18. Select the **CORNER RECTANGLE** tool.

19.

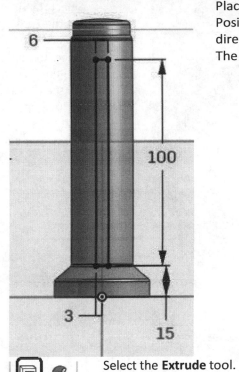

Place a corner point rectangle.
Position it centered on the origin in the width direction and 15 mm above the origin.
The rectangle is 6 mm wide and 100 mm high.

20.

Select the **Extrude** tool.

21.

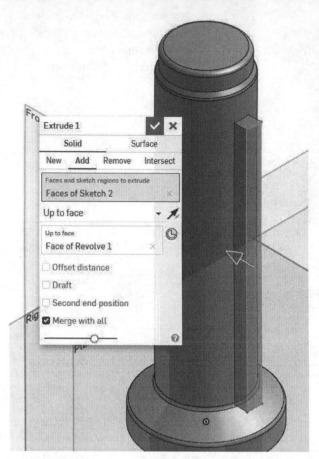

Select the **Add** option.
Set the Termination to **Up to Face**.
Select the body surface.
Use the Direction Arrow to flip the direction of the extrude into the part.
Enable **Merge with all.**
Click the Green check.

22.

Highlight Extrude1 in the browser.
Right click and select **Rename**.

23.

Rename **Grip.**

24.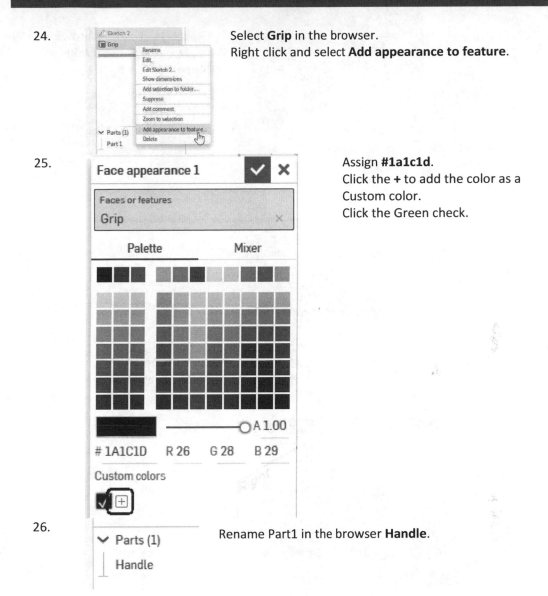

Select **Grip** in the browser.
Right click and select **Add appearance to feature**.

25.

Assign **#1a1c1d**.
Click the **+** to add the color as a Custom color.
Click the Green check.

26.

Rename Part1 in the browser **Handle**.

27.

Select **Handle** in the browser. Right click and select **Edit Appearance**.

28.

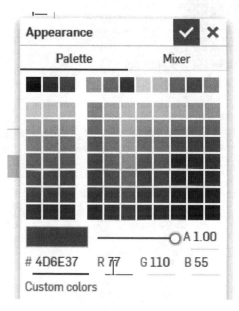

Assign **#4d6e37**.
Click the Green check.

29.   Select the **FILLET** tool.

30.

Add a **1 mm** fillet to the top face of the grip.

Notice that the fillet displays the color of the Handle/Part and not the color of the grip.

31.

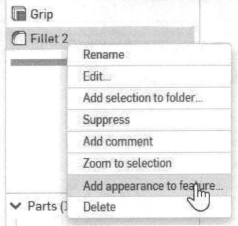

Highlight the Grip Fillet in the browser.

Right click and select **Add appearance to feature.**

32.

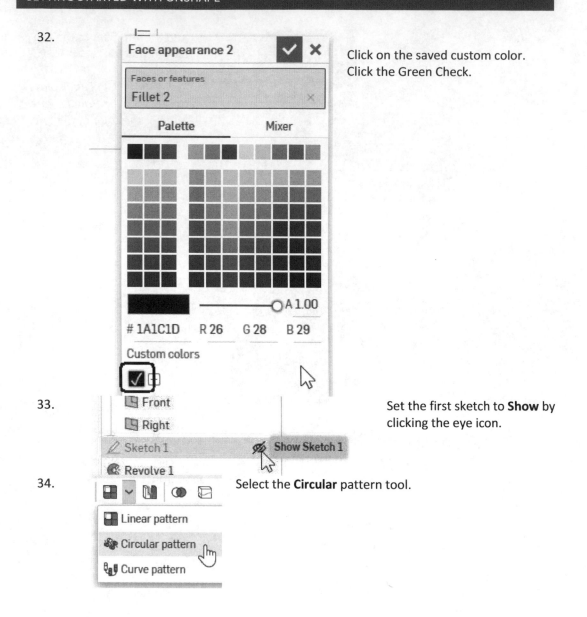

Click on the saved custom color.
Click the Green Check.

33.

Set the first sketch to **Show** by clicking the eye icon.

34.

Select the **Circular** pattern tool.

35.

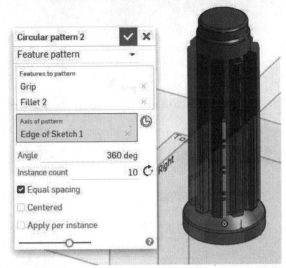

Select **Feature pattern** from the drop-down list.
Select the **Grip** and **Grip Fillet** to be patterned.
Select the vertical construction line in the first sketch to use as the pattern axis.
Set the termination to **360 deg.**
Set the Number of Instances to **10.**
Enable **Equal Spacing**.
Click the Green check.

36.

Set Sketch 1 to be hidden by clicking on the eye icon.

37.

Inspect your part and close the document.

Notes:

Chapter 6: Aluminum Hanger Project

Estimated Time: 45 minutes
Objectives:

- Sheet Metal model
- Flange
- Tab
- Hem
- Extrude
- Flat Pattern
- Assign material

1.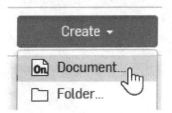

   Log in to your Onshape account.

   Click **Create→Document**.

2. **New document**

   Document name

   Aluminum Hanger

   Type **Aluminum Hanger**.

   Click **Create**.

3.

   Select **Workspace Units**.

4.

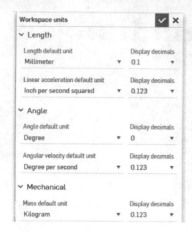

Set the Length to **Millimeter.**
Set the Length display decimals to **0.1.**
Set the Angle display decimals to **0.**
Set the Mass to **Kilogram.**

Click the Green check.

5.

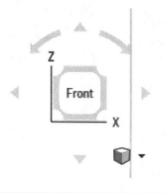

Select the **Front** plane.

Right click and select **New Sketch**.

6.

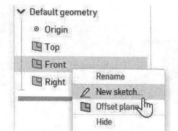

Switch to a **FRONT** view orientation.

7.

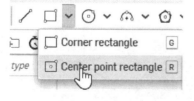

Select the **Center point rectangle**.

8.

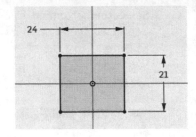

Add a 24 mm horizontal dimension and a 21 mm vertical dimension.

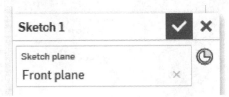

Exit the sketch.

9.   Select the **Sheet Metal Model** tool.

10.

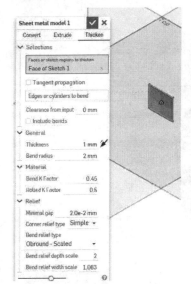

Enable **Thicken**.

Click inside the rectangle.

Set the Thickness to **1 mm**.

Set the Bend Radius to **2 mm**.

Accept the default values for the remainder of the prompts.

Green check.

11.

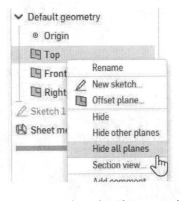

Select the Top plane in the browser.

Right click and select **Hide all planes**.

12.   Select the **Flange** tool.

13.

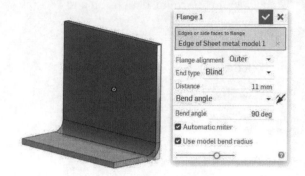

Select the bottom edge.
Set the Flange alignment to **Outer**.
Set the End type to **Blind**.
Set the Distance to **11 mm**.
Set the Bend angle to **90 deg**.
Enable **Automatic miter**.
Enable **Use model bend radius.**
Green check.

14.

Select the top of the flange for a new sketch.

15.

Switch to a **Top** view orientation.

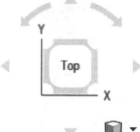

16.

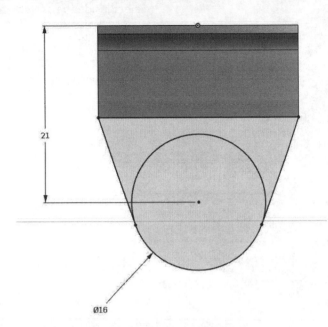

Ø16

Draw three lines and a circle (two angled lines that are tangent to the circle. One horizontal line that is collinear to the edge of the flange.)

Set the circle so it is aligned vertically to the origin.

Set the distance between the origin and circle centerpoint to 21 mm.

Set the circle diameter to 16 mm.

17.

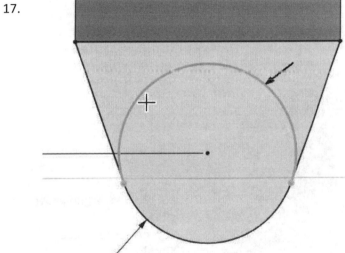

Use the **TRIM** tool to trim out the inner section of the circle.

Sketch 2	✓	✗
Sketch plane		
Face of Flange 1		x

Exit the sketch by clicking the Green check.

18. Select the **Tab** tool.

19.

Select inside the sketch profile.
Green check.

20. Select the **Hem** tool.

21.

Select the back edge of the top of the part.
Set the Hem type to **Straight**.
Disable **Flattened**.
Set the Inner Radius to **0.5 mm**.
Set the Total Length to **6 mm**.
Set the Hem alignment to **Outer**.
Set the Corner type to **Simple**.
Green check.

22.

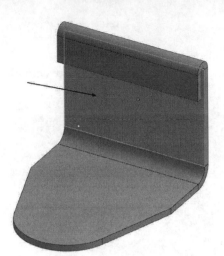

Select the front vertical face for a new sketch.

23.

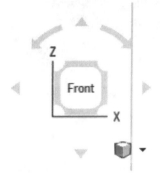

Switch to a **FRONT** view orientation.

24.

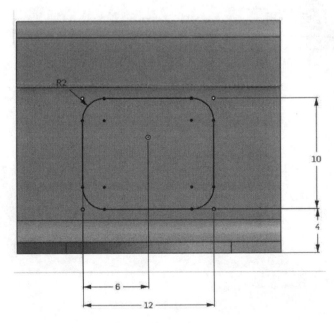

Place a corner rectangle.

Add R2 mm fillets to all four corners.

Dimension the rectangle as 12 mm in the horizontal and center.

Dimension the rectangle 10 mm in the vertical direction and locate 4 mm from the bottom edge.

25. Select the **Extrude** tool.

26.

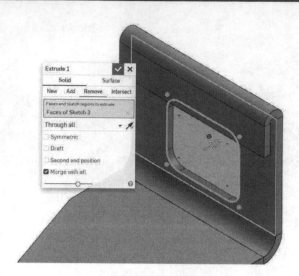

Enable **Remove**.
Set the Distance to
Through All.
Enable **Merge with all.**
Green check.

27.

Rename the extrude **window**.

🖐 Hem 1

✎ Sketch 3

📖 window

28.

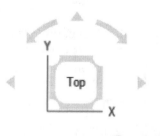

Select the top of the flange face for a new
sketch.

29.

Switch to a **Top** view orientation.

30.

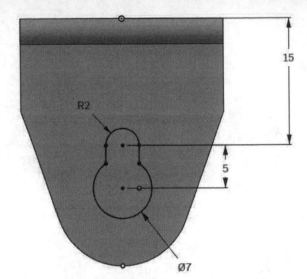

Sketch a keyhole.

31. Select the **Extrude** tool.

32.

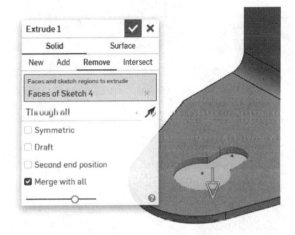

Enable **Remove**.
Set the Distance to
Through All.
Enable **Merge with all.**
Green check.

33. keyhole   Rename the extrude **keyhole**.

34. ∨ Parts (1)   Rename the Part **Hanger**.

   Hanger

35.

Right click on the part and **Assign material.**

36.

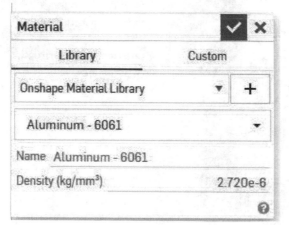

Assign **Aluminum 6061** to the part.

Green check.

37.

Click the **Flat Pattern** tab on the right side of the display window.

38.  Review the flat pattern.

Close the panel by clicking the tab.

39.

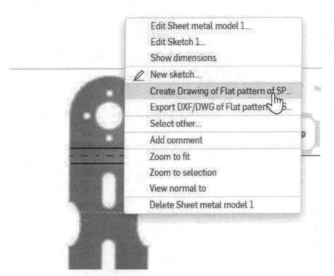

To create a drawing of the flat pattern, right click in the flat pattern panel and select **Create drawing of Flat Pattern**.

40.

onshape

Close the document.

Extra: Additional Projects

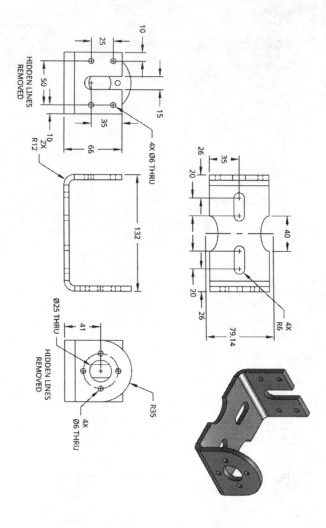

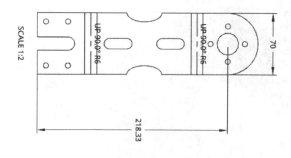

EX6-1

SPACER

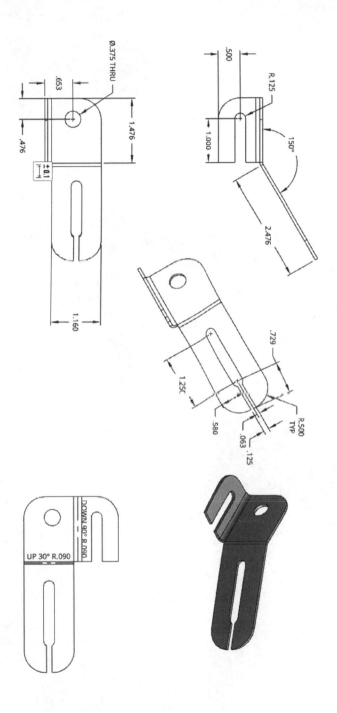

EX6-2
MOUNTING CLIP

Notes:

Chapter 7: Assemblies

Assembly 1: Handlebar and Grip Assembly

Estimated Time: 20 minutes
Objectives:

- Create an Assembly
- Fix Parts
- Add Mates

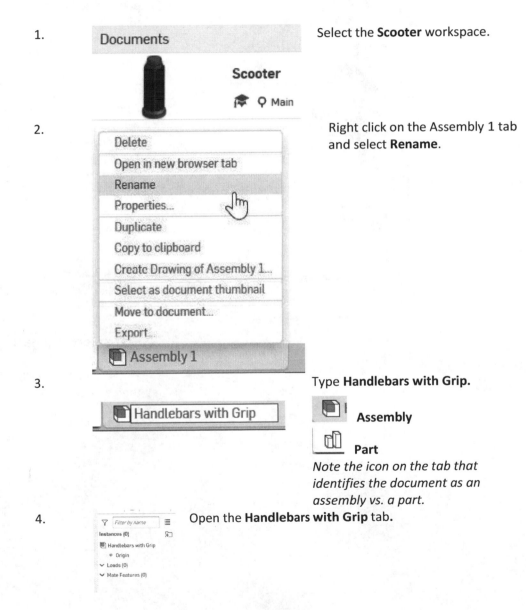

1. Select the **Scooter** workspace.

2. Right click on the Assembly 1 tab and select **Rename**.

3. Type **Handlebars with Grip.**

   Assembly

   Part

   Note the icon on the tab that identifies the document as an assembly vs. a part.

4. Open the **Handlebars with Grip** tab.

5.  Select **Insert parts/assemblies** from the ribbon.

6.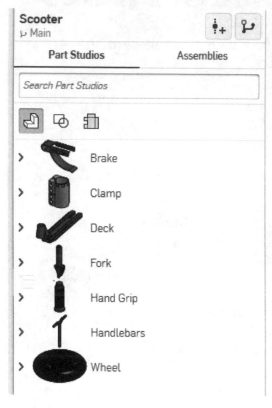

    A list of parts and assemblies available in the document will be displayed.

    The Parts that appear below the Part Studio Document are basically the parts listed in the browser for that component.

    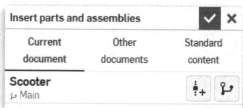

    If you collapse the list, you will see just the names of the components listed.
    To collapse, select the down arrows.

7. **Insert parts and assemblies** ✔ ✖

Current document	Other documents	Standard content

    Scooter
    ⌐ Main

    Left click on the Handlebars and they will automatically be inserted into the assembly.
    Green check to close the dialog box.
    This will automatically align the handlebars to the origin.

8.

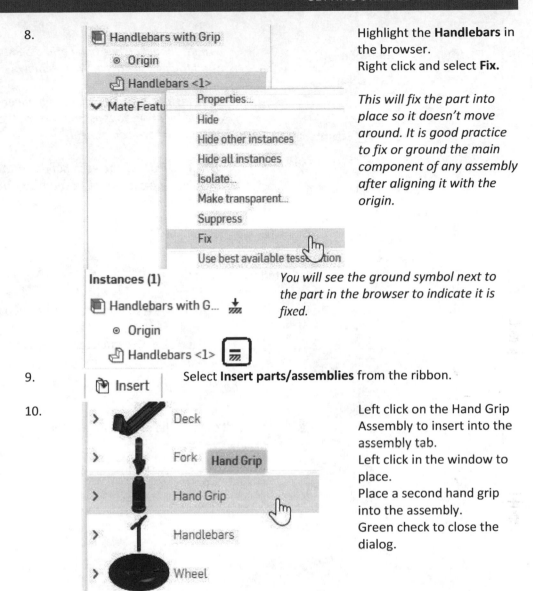

Highlight the **Handlebars** in the browser.
Right click and select **Fix.**

This will fix the part into place so it doesn't move around. It is good practice to fix or ground the main component of any assembly after aligning it with the origin.

You will see the ground symbol next to the part in the browser to indicate it is fixed.

9. Select **Insert parts/assemblies** from the ribbon.

10. Left click on the Hand Grip Assembly to insert into the assembly tab.
Left click in the window to place.
Place a second hand grip into the assembly.
Green check to close the dialog.

11.

Instances (3)

📄 Handlebars wit... ⬇

　⊛ Origin

　📁 Handlebars ...

　📁 Handle <1> ⅄

　📁 Handle <2> ⅄

There should be two hand grip assemblies and the handlebars in the assembly.

The Instances counts the total number of components in your assembly – not the count for each component.

The UCS icon next to each handle instance displays the degrees of freedom available on the part.

12.

Enable **Snap Mode**.

13.

Select the **Fastened** mate tool.

14.

Zoom into the handlebar. You will see a small mate connector as you hover over the bar.

15.

Left click on the center circle at the end of the handlebar.

16.

Zoom into the hand grip. You will see a small mate connector as you hover over the center of the bottom cylinder.
Left click on the center circle.

17.

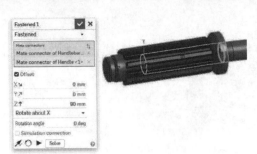

Enable **Offset**.
Note that the colors for X, Y, and Z correlate with the axis colors on the mate connector.
Enter **-90 mm** in the Z field. This adds a distance of 90 mm so that the grip is mounted onto the handle-bar.

18.

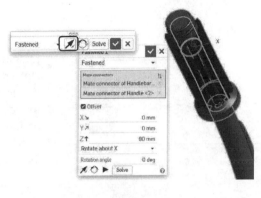

Reverse the direction if necessary.

If you need to switch the direction, you need to change the Offset value to **+90 mm**.

Click the Green check.

19.

Hold down your left mouse button over the grip to try to move it.

It is fastened into place.

20.

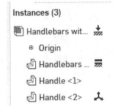

Check in the browser and you will see the degrees of freedom symbol is no longer next to the Handle part.

21. Select the **Fastened** mate tool.

22. Enable **Snap Mode**.

23.

Zoom into the handlebar. You will see a small mate connector as you hover over the bar.
Left click on the center circle at the end of the handlebar.

24.

Left click on the center circle at the bottom of the Hand Grip.

25.

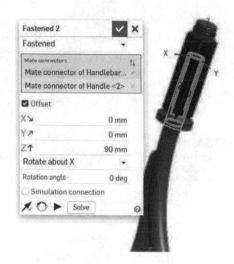

Enable **Offset**.
Note that the colors for X, Y, and Z correlate with the axis colors on the mate connector.
Enter **90 mm** in the Z field. This adds a distance of 90 mm so that the grip is mounted onto the handlebar.
Reverse the direction if necessary.

Click the Green check.

26.

Inspect your completed
assembly.
Close the document.

Assembly 2: Scooter Assembly

Estimated Time: 60 minutes
Objectives:
- Create an Assembly
- Add Mates

1.

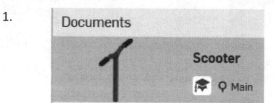

Select the **Scooter** workspace.

2.

Select the **+** tab.
Select **Create assembly**.

3.

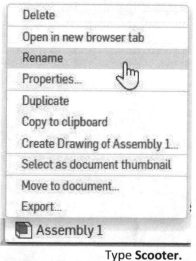

Right click on the Assembly tab and select **Rename**.

4.

Type **Scooter.**

5.   Insert   Select **Insert parts/assemblies** from the ribbon.

6.   Left click to select the **Deck** component.

Close the dialog to align the origin of the deck with the assembly origin.

7.   Locate the part in the browser.
Right click and select **Fix**.

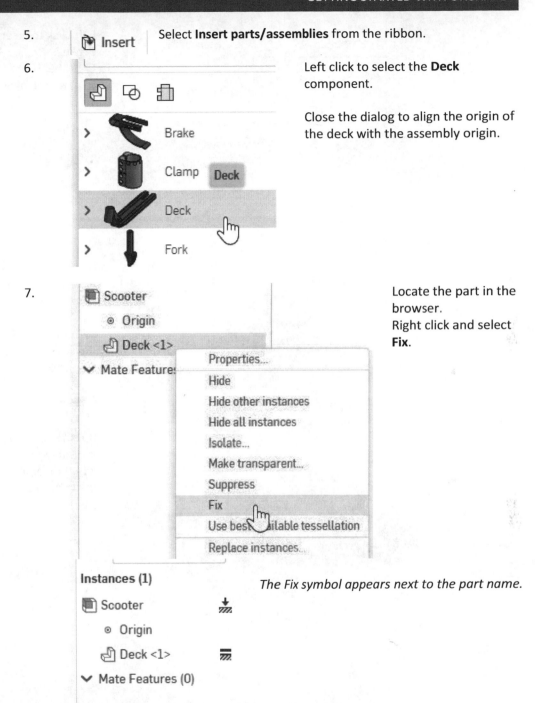

The Fix symbol appears next to the part name.

8.   Insert   Select **Insert parts/assemblies** from the ribbon.

9.

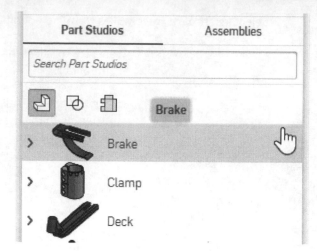

Left click to insert the **Brake**.
Click the Green check to close the dialog.

10.

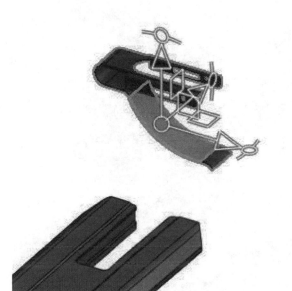

Left click on the brake to activate the Triad Manipulator.
Use the Triad Manipulator to orient the wheel for easier assembly.

Be sure the Snap Mode is disabled or you won't be able to activate the Triad Manipulator.

11.

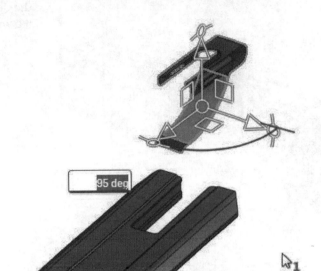

To rotate, simply hold
the left mouse down
on the circle symbol
next to the arrowhead
and drag.

12.

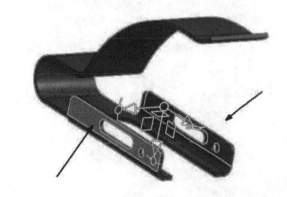

Select the outside
faces of the Brake.

13.

*In the lower right corner of the display
window, you will see the distance
between the selected faces is
displayed as **42 mm**.*

Release the selection by left clicking in
the window.

14.

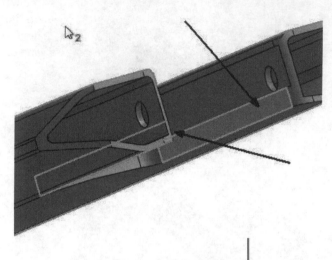

Select the inside faces of the deck where the brake will mount.

15.

Parallel Dist: **42.0** mm

The opening is displayed as 42 mm.

16. Select the **Fastened** mate tool.

17.

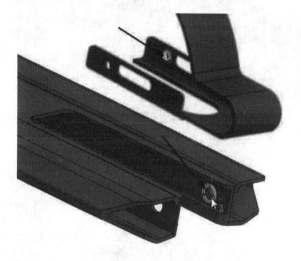

Select the hole center on the deck and the hole center on the brake.
Click the Green check.

18.

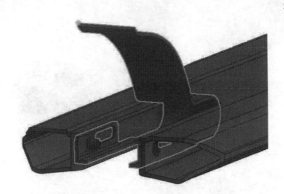

Inspect the assembly so far.

19. 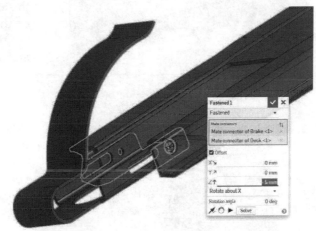 Select the **Fastened** mate tool.

20.

Select the hole center on the deck and the hole center on the brake.
Enable Offset.
Set the value to **-.5 mm.**
Click the Green check.

21. | 📖 Insert | Select **Insert parts/assemblies** from the ribbon.

22.

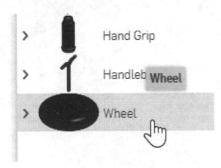

Use the scroll bar to scroll down to the **wheel.**
Insert two wheels into the assembly.
Click the Green check to close the dialog.

23.

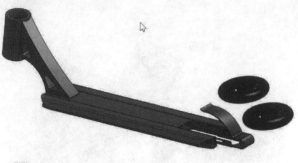

There should be a Deck (fixed), brake, and two wheels in the assembly.

24.

Use the Triad Manipulator to orient the wheel for easier assembly.
To rotate, simply hold the left mouse down on the circle symbol next to the arrowhead and drag.

25. Select the **Cylindrical** mate.

26.

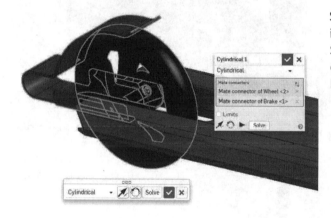

Select the center hole in the wheel.
Select the corresponding inside hole in the brake.

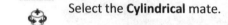

27.

Verify that you are placing a Cylindrical mate.
Click the Green check.

28.

Place the left mouse over the wheel.
Hold down the left mouse and rotate the wheel.

Notice the wheel rotates, but it also moves side to side.

29.

Use the triad tool to move the wheel outside the assembly.

Notice if you try to move it using the wrong axis, it won't move due to the existing mate.

30.

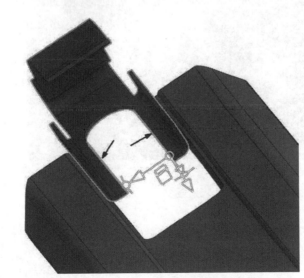

Measure the brake opening by selecting the two faces/edges.

Parallel Dist: **24.0** mm

You should get a distance of 24 mm.

31.

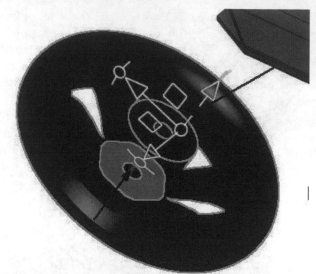

Measure the width of the wheel.

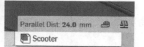

You should get a distance of 24 mm.

32. Select **PLANAR** mate.

33.

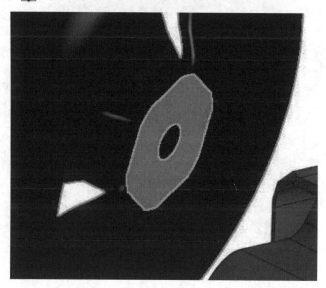

Select the outside face of the wheel.

34.

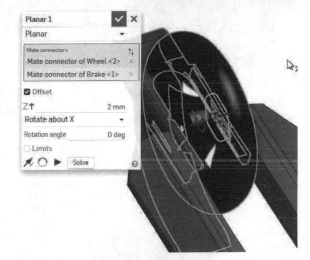

Select the face of the ledge on the brake.

35.

Green check.

36.

You should be able to rotate the wheel, but it should maintain its position centered in the deck.

37. Select the **Insert** tool.

38.

Left click on the **Fork** to add to the assembly. Select the Green check to close the dialog.

39.   Select the **Cylindrical** mate tool.

40.

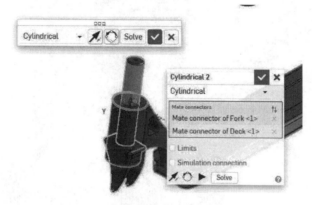

Select the two cylindrical faces of the fork and the deck. Select the Green check.

41.   Select the **Planar** mate tool.

42.

Select the flat horizontal face on the fork.

43.

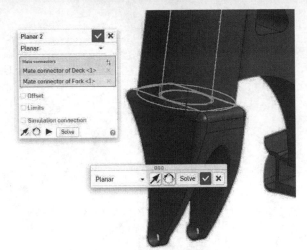

Select the bottom of the cylinder on the deck.
Click the Green check.

44. Select the **Cylindrical** mate tool.

45.

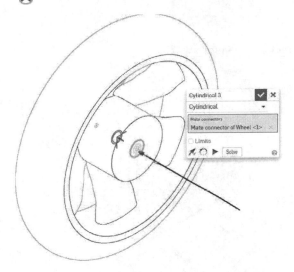

Select the hole on the wheel.

46.

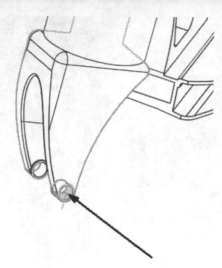

Select the hole on the bottom of the fork.

Use the flip arrow tool if necessary to change the orientation of the wheel.
Click the Green check.

47.   Select the **Planar** mate tool.

48.

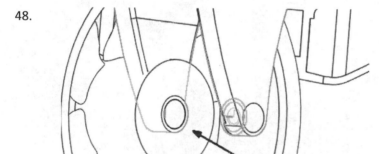

Select the flat horizontal face on the wheel.

49.

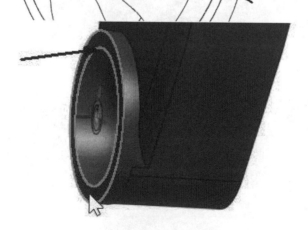

Select the flat inside face of the fork's boss. Select the Green check.

50.

This is the assembly so far.
Check to see which parts can move.
You should be able to rotate the wheels and turn the fork.

51. 🖢 Insert   Select the **Insert** tool.

52.

Left click on the **Assemblies** link. Left click on the **Handlebars with Grips** to insert into the assembly.
Select the Green check to close the dialog.

53.

Orient the handlebars so they are perpendicular to the deck.

54. ⬡ Select the **Fastened** tool.
By selecting the Fastened tool, we can align the handlebar assembly with the fork so they move together.

55.

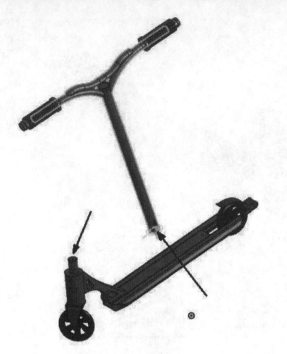

Select the two faces indicated on the handlebars and fork.

56.

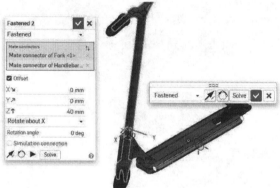

Enable **Offset**.
Set the Z value to **40 mm**.
Select the Green check.

57.

Try moving the handlebars and notice that the wheel and fork turn with the handlebars.

58. Insert   Select the **Insert** tool.

59.

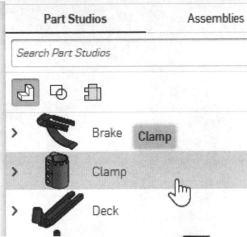

Select the **Clamp**.
Left click to place into the window.
Select the Green check to close the dialog.

60.

Orient the clamp so the holes are parallel to the deck.

61.

Add a **cylindrical** mate between the clamp and the shaft of the fork.

62.

Add a **Planar** mate between the bottom of the clamp and the top of the fork.

63. Select **PARALLEL** mate.

64.

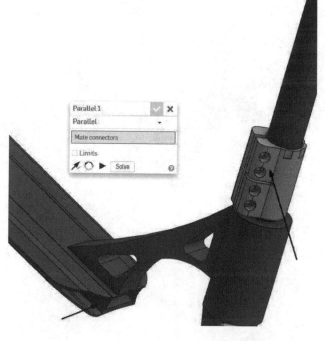

Select the flat face of the clamp.

Select the vertical wall of the deck.

Green check.

65.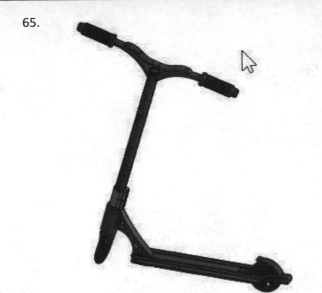

onshape

Close the document.

Assembly 3: Copy and Paste Public Documents

Estimated Time: 30 minutes

Objectives:

- Search Public documents
- Add a public document to an assembly
- Change Properties
- Assign Material
- Assign Appearance
- Add Fastened Mates

1.

Select the **Scooter** workspace.

2.

You should be in the Scooter assembly.
If not, click on the assembly tab.

3.   Insert

Select the **INSERT** tool.

4.

Click **Other documents**.

Click **Public**.

5.

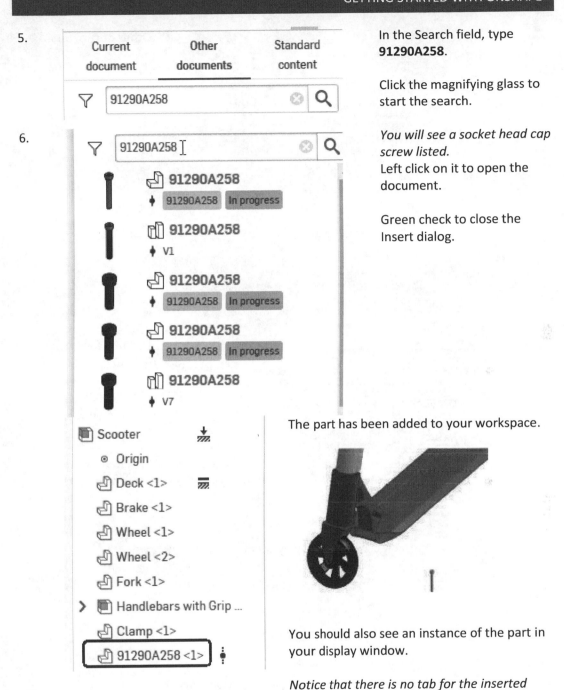

Current document	Other documents	Standard content

| ▽ | 91290A258 | ⊗ | 🔍 |

In the Search field, type **91290A258**.

Click the magnifying glass to start the search.

6.

| ▽ | 91290A258 | ⊗ | 🔍 |

91290A258
  ♦ 91290A258   In progress

91290A258
  ♦ V1

91290A258
  ♦ 91290A258   In progress

91290A258
  ♦ 91290A258   In progress

91290A258
  ♦ V7

You will see a socket head cap screw listed.
Left click on it to open the document.

Green check to close the Insert dialog.

📄 Scooter   ⊥

  ⊙ Origin

  Deck <1>   🔲

  Brake <1>

  Wheel <1>

  Wheel <2>

  Fork <1>

> 📦 Handlebars with Grip ...

  Clamp <1>

  91290A258 <1>   ⋮

The part has been added to your workspace.

You should also see an instance of the part in your display window.

Notice that there is no tab for the inserted part.

7.   📋 Insert   Select the **INSERT** tool.

8.

Insert parts and assemblies   ✓ ✗

| Current document | Other documents | Standard content |

▽ *Search or paste URL* 🔍

👤 My Onshape
🕐 Recently opened
📄 Created by me
👤 Shared with me
🌐 Public

Click **Other documents**.

Click **Public**.

9.

| Current document | Other documents | Standard content |

▽ 92497A300 ⊗ 🔍

❮ **Public**

In the Search field, type **92497A300**.

Click the magnifying glass to start the search.

10.

92497A300

92497A300

You will see a METRIC HEX NUT listed. Left click on it to open the document.

11.

Left click in the display window to place the part.

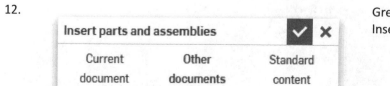

1 ❯ 📋 Handlebars with Grip ...
    👍 Clamp <1>
    👍 91290A258 <1> ⋮
    👍 92497A300 <1> ⋮

The part has been added to the workspace.

12.

Insert parts and assemblies   ✓ ✗

| Current document | Other documents | Standard content |

Green check to close the Insert dialog.

13.

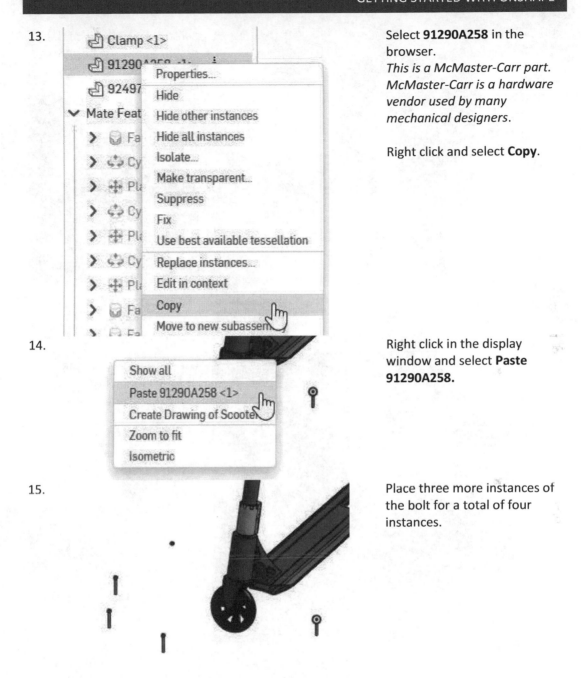

Select **91290A258** in the browser.
This is a McMaster-Carr part. McMaster-Carr is a hardware vendor used by many mechanical designers.

Right click and select **Copy**.

14.

Right click in the display window and select **Paste 91290A258.**

15.

Place three more instances of the bolt for a total of four instances.

16. Select the **Fasten** mate.

17.

Select the underside of the bolt head and the bottom of the counterbore on the clamp.

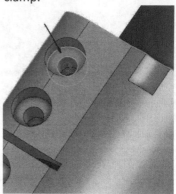

18.

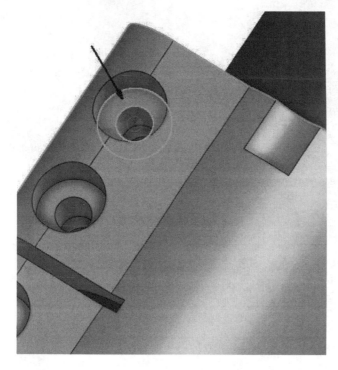

Use the arrow to flip the direction of the insert, if necessary.

Click the Green check.

19.

Repeat to fill the remaining counterbore holes with the bolts.

Close the Fastened dialog.

Fastened ▾ Solve ✓ ✗

20.

> 📄 Handlebars with Gr...
🔩 Clamp <1>
🔩 91290A258 ...
🔩 92497A300
🔩 91290A258
🔩 91290A258
🔩 91290A258
⌄ Mate Features (
> Fastened
> Cylindric
> Planar 1
> Cylindric
> Planar 2
> Cylindric
> Planar 3
> Fastened

Properties
Hide
Hide other instances
Hide all instances
Isolate...
Make transparent...
Suppress
Fix
Use best available tessellation
Replace instances...
Edit in context
Copy
Move to new subassem
Update linked document...

Highlight **92497A300** in the browser.
This is also a McMaster-Carr part.

Right click and select **Copy.**

21.

Fix
Use best available tessellation
Copy 92497A300 <1>
Paste 92497A300 <1>
Create Drawing of 92497A300...
Create Drawing of Scooter...

Right click and select **Paste 92497A300.**

22.

Place three instances of the nut.

23.   Select the **Fasten** mate.

24.

Select the underside of the nut and the bottom of the bolt on the clamp.

25.

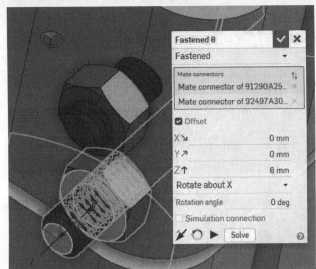

Enable **Offset**.
Set the Z offset to **6 mm**.
Click the Green check.

Change the direction if required.

26.

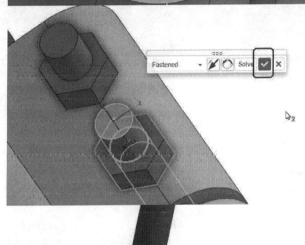

Repeat to mount the remaining nuts to the bolts.

If you green check the small toolbar, the Fastened tool remains open so you can place multiple mates.

27.

Close the document.

Assembly 4: Import SOLIDWORKS Parts

Estimated Time: 30 minutes
Objectives:

- Importing a SOLIDWORKS Part
- Change Properties
- Assign Material
- Assign Appearance
- Add Fastened Mates

1.

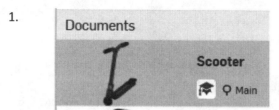

Select the **Scooter** workspace.

2.

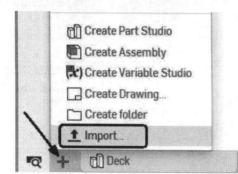

Select the **+** tab.
Select **Import**.

3.

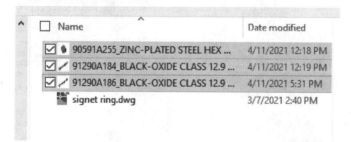

Locate the
91290A184.sldprt
90591A255.sldprt
91290A186.sldprt
files.

*These can be
downloaded from
sdcpublications.com or
mcmaster.com. These
are SOLIDWORKS files
for a cap screw and nut
from McMaster-Carr.
Notice that you can
import more than one
file at a time by
holding down the Ctrl
key or clicking in the
check box.*

Click **Open**.
Enable **Create a composite part**...this creates a single part and combines all the features.

4.

☑ Create a composite part when importing multiple or non-solid bodies ?

Import   Cancel

Click **Import**.

Onshape's coordinate system has the Z axis in the up direction and the XY axis forming the top plane. If the imported models are oriented with the Y axis up and the XZ axis forming the top plane, the models may need to be re-oriented.

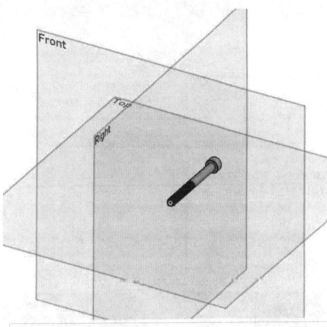

5.

Notifications (2 unread, 2 total)   mark all as read   delete all ✕

⌄ **Upload in progress**   ●
    Uploading 91290A184_BLACK-OXIDE CLASS 12.9 SOCKET HEAD CAP
    SCREW.SLDPRT.
    Importing...
    Created at 12:23 PM Today

⌄ **Translation completed**   ●
    90591A255_ZINC-PLATED STEEL HEX NUT.SLDPRT was translated successfully.
    Created at 12:23 PM Today

A window will appear indicating the files have been imported. Left click on the X in the upper right corner to close the window.

6.

📄 91290A184_BLACK-OX...   📄 90591A255_ZINC-PLA...   📁 CAD Imports

Use the arrow keys on the bottom scroll bar to see the new tabs.

You can re-organize the tabs by dragging and dropping them into the desired location.

There is a folder for CAD imports.
The folder contains one tab for each native SOLIDWORKS document imported.
Each imported file is converted to an Onshape part.

7.

If you click on the CAD imports tab, it expands to show tabs for each imported SOLIDWORKS part.

8.

These tabs show information about the part file but no geometry.

9.

Right click on the tabs with the *.sldprt* extensions.

If you delete the imported tabs, you will delete the Onshape corresponding tabs.

10.

Click on the **Home** tab to return to the work environment.

11.

Select the **90591A255** tab.

12.

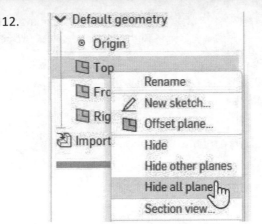

In the browser, select the **Top** plane. Right click and select **Hide all planes**.

You see a nut.

13.

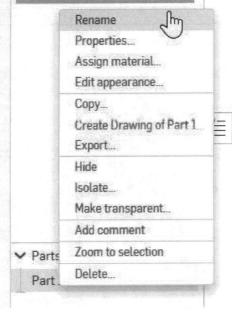

Right click on the Part in the browser. Select **Rename.**

14.

✓ Parts (1)

90591A255

Rename with the part number **90591A255.**

15.

Rename
Properties...
Assign material...
Edit appearance...
Copy...
Create Drawing of 90591A2!
Export...
Hide
Isolate...
Make transparent...
Add comment
Zoom to selection
Delete...

∨ Parts (1)

90591A2

Highlight the part name in the browser. Right click and select **Properties.**

16.

Properties

90591A255

Name *

90591A255

Description

HEX NUT, M4 STEEL

Category

Onshape Part

Part number

90591A255

Type: **HEX NUT, M4 STEEL** in the Description field. Type: **90591A155** in the Part Number field. Click **Apply** and **Close.**

The information in Properties can be used in bills of materials or parts lists.

17.

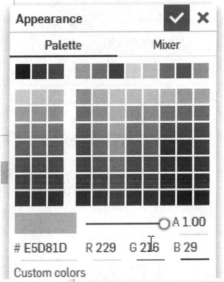

Highlight the part name in the browser. Right click and select **Edit appearance.**

18.

Assign the following values:
R: **229**,
G: **216**,
B: **29**.
Select the Green check.

19.

Highlight the part name in the browser. Right click and select **Assign material.**

20.

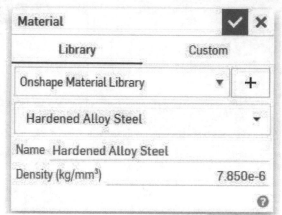

Set the Material to **Hardened Alloy Steel** using the drop-down list.
Click the Green check.

21.

Select the **91290A184** tab.

You see a bolt.

22.

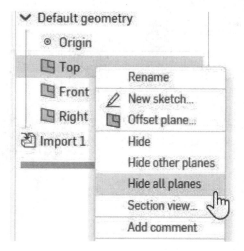

In the browser, select the **Top** plane.
Right click and select **Hide all planes**.

23.

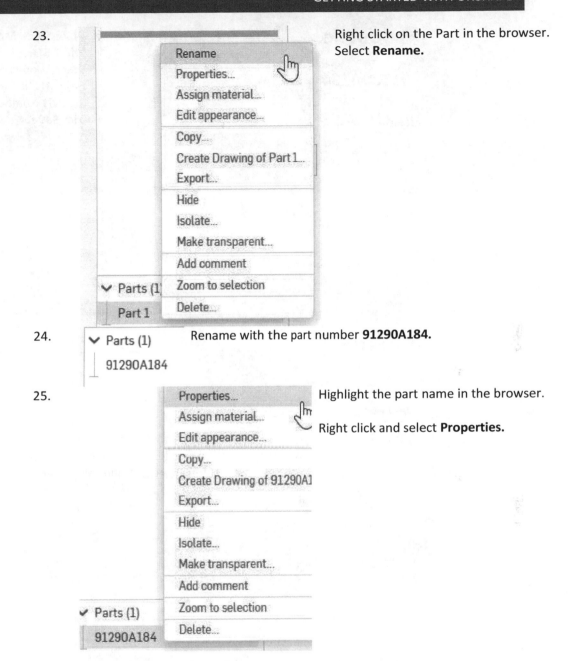

Right click on the Part in the browser.
Select **Rename.**

24.   ∨ Parts (1)   Rename with the part number **91290A184.**

      91290A184

25.   Highlight the part name in the browser.

Right click and select **Properties.**

26.

Properties

91290A184

Name *

91290A184

Description

SHCS, M4 X 40 MM L STEEL

Category

Onshape Part

Part number

91290A184

Revision

Type: **SHCS, M4 X 40 MM L STEEL** in the Description field.
Type: **91290A184** in the Part Number field.
Click **Save**.

27.

| Edit appearance... |
| Copy... |
| Create Drawing of 91290A18 |
| Export... |
| Hide |
| Isolate... |
| Make transparent... |
| Add comment |
| Zoom to selection |
| Delete... |

∨ Parts (1)
  91290A18

Highlight the part name in the browser.
Right click and select **Edit appearance.**

28.

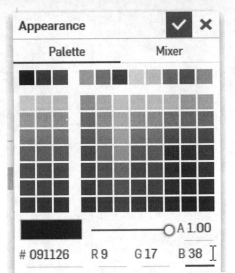

Assign the following values:
R: **9**,
G: **17**,
B: **38**.
Click the Green check.

29.

Highlight the part name in the browser. Right click and select **Assign material.**

30.

Set the Material to **300 Series Stainless Steel** using the drop-down list.

If you type the material name in the search box, all materials with those letters will appear.

Green check.

31.

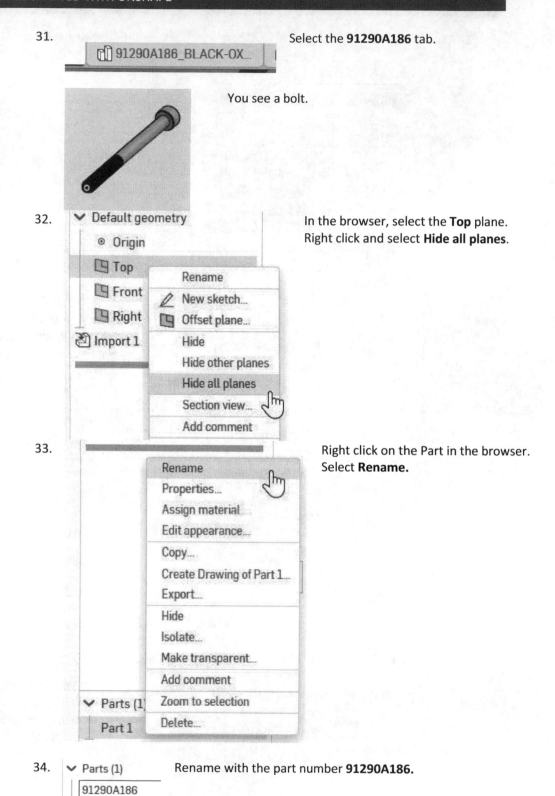

Select the **91290A186** tab.

You see a bolt.

32. In the browser, select the **Top** plane. Right click and select **Hide all planes**.

33. Right click on the Part in the browser. Select **Rename.**

34. Rename with the part number **91290A186.**

35.

Properties...

Assign mater̶

Edit appearance...

Copy...

Create Drawing of 91290A

Export...

Hide

Isolate...

Make transparent...

Add comment

Zoom to selection

⌄ Parts

Delete...

91290A186

Highlight the part name in the browser.

Right click and select **Properties.**

36.

Properties

91290A186

Name *

91290A186

Description

SHCS, M4 X 50 MM L STEEL

Category

Onshape Part

Part number

91290A186

Type: **SHCS, M4 X 50 MM L STEEL** in the Description field.
Type: **91290A186** in the Part Number field.
Click **Save**.

37.

| Assign material... |
| Edit appearance... |
| Copy... |
| Create Drawing of 91290A186... |
| Export... |
| Hide |
| Isolate... |
| Make transparent... |
| Add comment |
| Zoom to selection |
| Delete... |

Parts (

91290/

Highlight the part name in the browser.
Right click and select **Edit appearance.**

38.

Appearance ✓ ✕

Palette   Mixer

◯ A 1.00

091126   R 9   G 17   B 38

Assign the following values:
R: **9**,
G: **17**,
B: **38**.
Click the Green check.

39.

| Properties... |
| Assign material... |
| Edit appearance... |
| Copy... |
| Create Drawing of 91290A186... |
| Export... |
| Hide |
| Isolate... |
| Make transparent... |
| Add comment |
| Zoom to selection |
| Delete... |

P

91290A186

Highlight the part name in the browser.
Right click and select **Assign material.**

40.

Material ✓ ✕

Library   Custom

Onshape Material Library ▼ **+**

Hardened Alloy Steel ▼

Name Hardened Alloy Steel

Density (kg/mm³) 7.850e-6

❓

Set the Material to **Hardened Alloy Steel** using the drop-down list.
Click the Green check.

41.

🗐 Scooter

Select the **Scooter** tab.

42.

🖱 Insert

Select the **Insert** tool.

43.

Insert parts and assemblies ✓ ✕

| Current document | Other documents | Standard content |

Scooter
ᵖ Main

Part Studios   Assemblies

Search Part Studios

⬇ ⬀ 🗐

91290A184_BLACK-OXIDE CLASS 12.9 SOCKET HEAD CAP SCREW

› ⟍ 91290A184_BLACK-OXIDE CL...

› ⟍ Brake

Select **91290A184.**
This is the SOLIDWORKS part you imported.

44.

Place one instance of the bolt.
Click the Green check to close the dialog.

45. Select the **Fasten** mate.

46.

Select the underside of
the bolt head.

47.

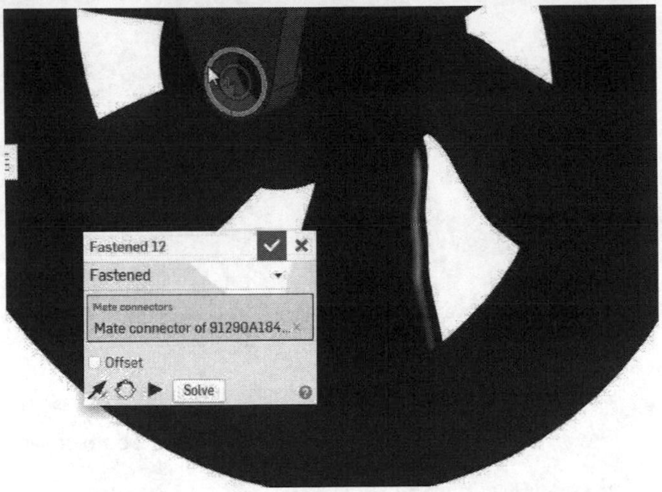

Fastened 12

Fastened

Mate connectors
Mate connector of 91290A184... ×

Offset

Solve

Select the hole on the left side of the fork. Use the flip arrow button if you need to flip the direction of the bolt.
 Click the Green check.

48.

onshape

Close the document.

Assembly 5: Troubleshoot an Assembly

Estimated Time: 10 minutes

Objectives:

- Isolate
- Edit Feature
- Hole
- Mirror
- Rollback bar

1.

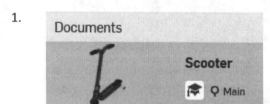

Select the **Scooter** workspace.

2.

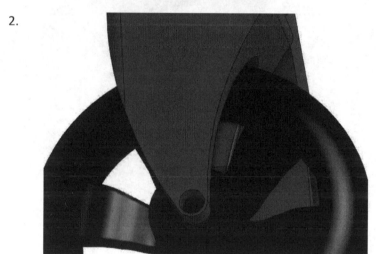

Zoom into the front wheel.

3.

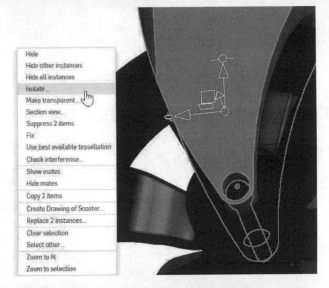

Select the fork and the cap screw mounted in the fork.

Right click and select **Isolate**.

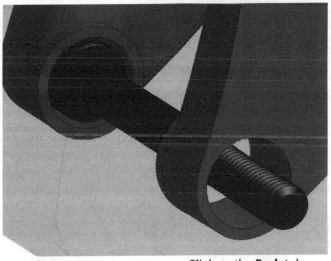

You can see that the hole is not correct.

4.

Click on the **Fork** tab.

5.

Select the hole in the display window.

The hole also highlights in the browser.

6.

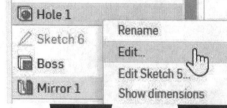

Highlight the Hole.

Right click and select **Edit**.

7.

Change the hole to a **Counterbore**.
Set the Termination to **Blind**.
Set the through hole to 4 mm.
Set the C'bore diameter to **8 mm**.
Set the C'bore depth to **4 mm**.
Set the Hole Depth to **10 mm**.

Green check.

8.

Drag the rollback bar below the Hole.

9. Select the **Mirror** tool.

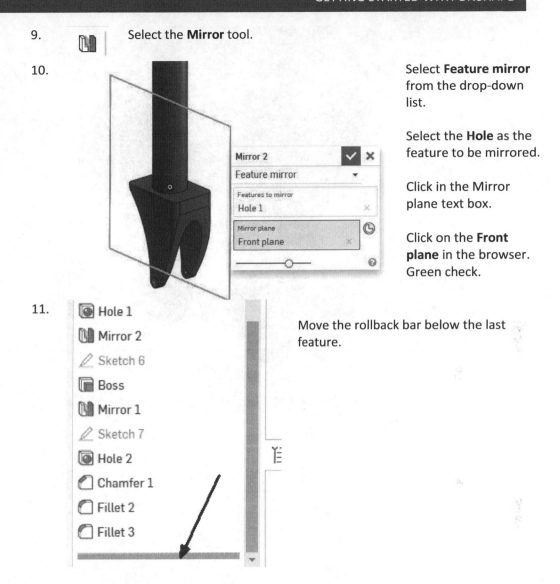

10. Select **Feature mirror** from the drop-down list.

Select the **Hole** as the feature to be mirrored.

Click in the Mirror plane text box.

Click on the **Front plane** in the browser. Green check.

11. Move the rollback bar below the last feature.

12.

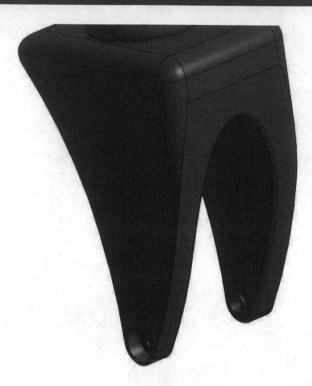

onshape

Close the document.

Assembly 6: Continue Scooter Assembly

Estimated Time: 20 minutes

Objectives:

- Edit mate
- Insert
- Fasten

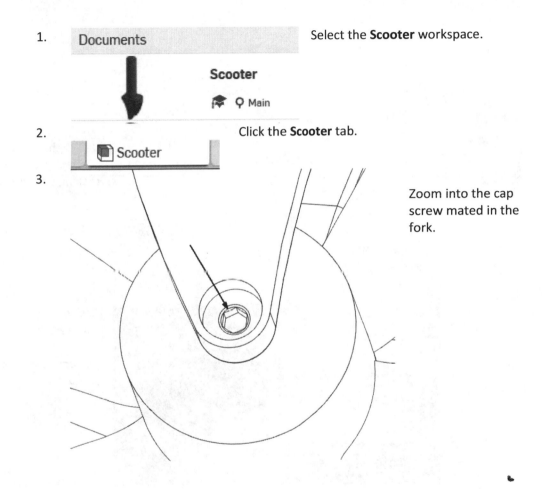

1. Select the **Scooter** workspace.

2. Click the **Scooter** tab.

3. Zoom into the cap screw mated in the fork.

4.

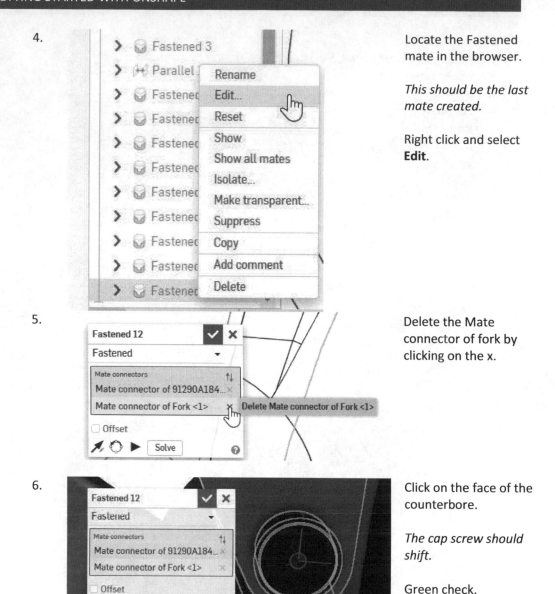

Locate the Fastened mate in the browser.

This should be the last mate created.

Right click and select **Edit**.

5.

Delete the Mate connector of fork by clicking on the x.

6.

Click on the face of the counterbore.

The cap screw should shift.

Green check.

7. Insert

Select the **Insert** tool.

8.

Select **90591A255.**
This is the SOLIDWORKS part you imported.

9.

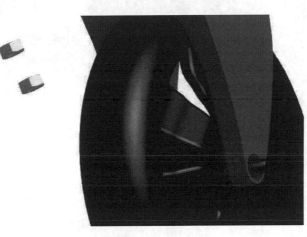

Place two instances of the nut.
Select the Green check to close the dialog.

10. Select the **Fasten** mate.

11.

Select the circular face of the nut.

12.

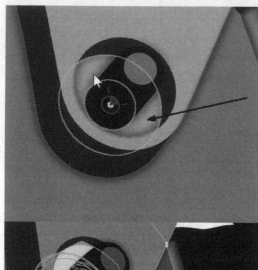

Select the bottom of the counterbore on the right side of the deck.

13.

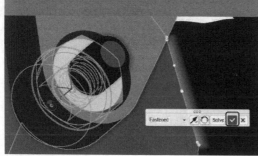

Use the flip arrow button if you need to flip the direction of the bolt.
Click the Green check on the small toolbar.

14.   📋 Insert   Select the **Insert** tool.

15.

Select **91290A186 cap screw.**
This is the SOLIDWORKS part you imported.

16.

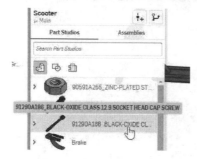

Place one instance in the display window.

Close the Insert panel.

17.   🛢   Select the **Fasten** mate.

18.

Select the underside of the head of the second 91290A186 cap screw.

19.

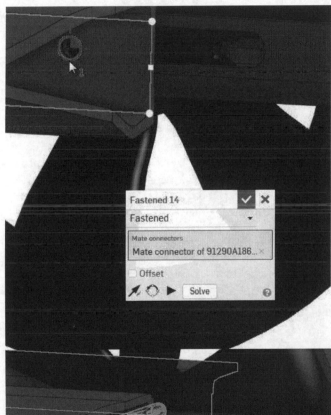

Select the hole in the rear of the deck.

20.

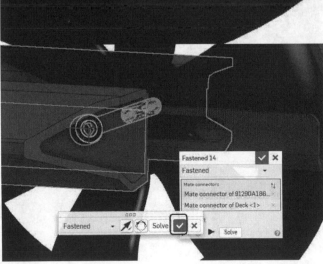

Reverse the direction, if needed.

Green check if the mate preview looks correct.

21.

Select the circular face of the nut.

Make sure the connector is in the center of the nut.

22.

Orbit the deck to the other side and select the outside hole on the rear of the deck.

23.

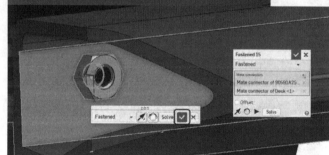

Use the flip arrow button if you need to flip the direction of the bolt.
Select the Green check.

24. Close the document.

GETTING STARTED WITH ONSHAPE

Assembly 7: Create Folder

Estimated Time: 10 minutes

Objectives:

- Organize Assembly
- Create Folder

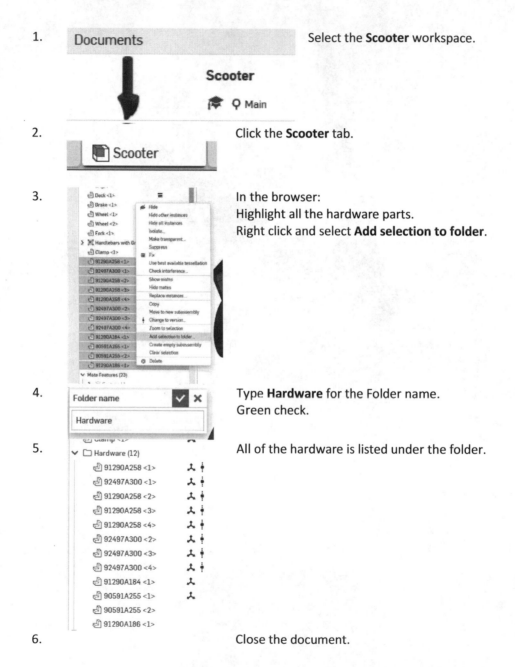

1. Select the **Scooter** workspace.

2. Click the **Scooter** tab.

3. In the browser:
   Highlight all the hardware parts.
   Right click and select **Add selection to folder**.

4. Type **Hardware** for the Folder name.
   Green check.

5. All of the hardware is listed under the folder.

6. Close the document.

Chapter 8: Drawings

Create a Template from a DWT file

Estimated Time: 5 minutes

Objectives:

- Import a DWT file

DWT files are drawing templates created using an Autodesk format.
You can import any dwt file into Onshape.

1.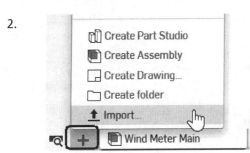

    Select the **Sample - Wind Meter - Copy** workspace.
    This workspace was created in Chapter One.

2.

    Select the Add + tab.
    Select **Import**.

3.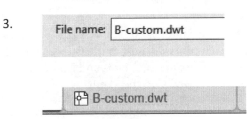

    Locate *B-custom.dwt.*
    This file can be downloaded from sdcpublications.com.
    Select **Open**.
    You see the tab with the import.

4.

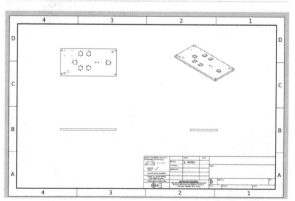

Locate the Base Plate in the Parts list in the browser.

Right click and select **Create drawing of Base Plate**.

5.

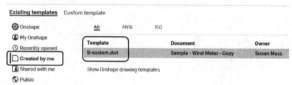

Left click on **Created by me**.
The B-custom.dwt template is listed as an available template. Highlight the *B-custom.dwt* template.

6.

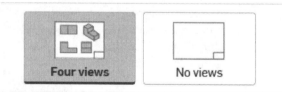

Enable **Four views**.
This will automatically place the front, top, right, and isometric views on the sheet.
Click **OK**.

7.

There is a slight pause as the drawing is generated.

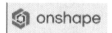

Close the document.

Create a Template from Scratch

Estimated Time: 15 minutes

Objectives:

- Create a template

1.  Select the **Sample - Wind Meter - Copy** workspace.
This workspace was created in Chapter One.

2.  Click on the **Wind Meter Main** tab to open.

3. 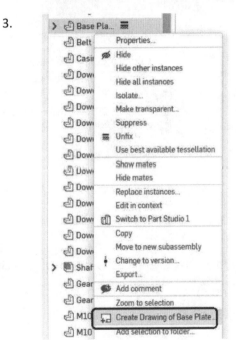 Locate the Base Plate in the browser.
Right click and select **Create drawing of Base Plate**.

4.  Existing templates   **Custom template** Select **Custom template**.

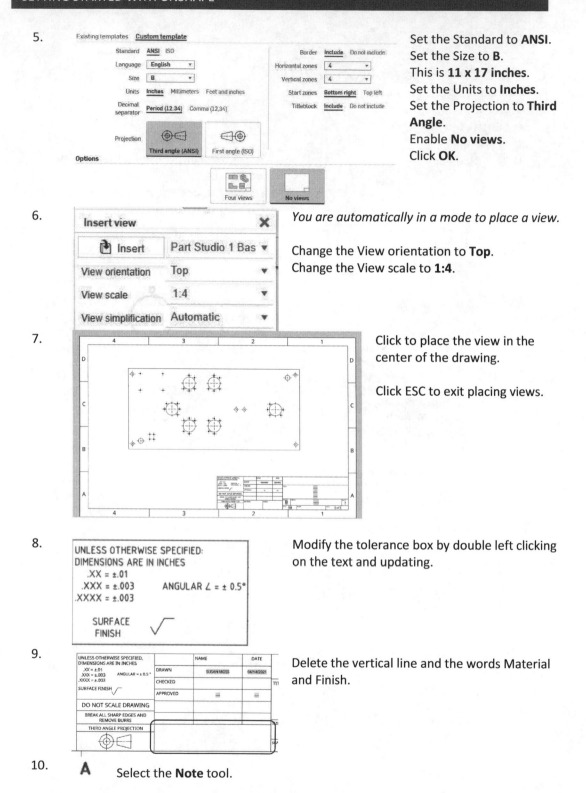

5.
Set the Standard to **ANSI**.
Set the Size to **B**.
This is **11 x 17 inches**.
Set the Units to **Inches**.
Set the Projection to **Third Angle**.
Enable **No views**.
Click **OK**.

6.
You are automatically in a mode to place a view.

Change the View orientation to **Top**.
Change the View scale to **1:4**.

7.
Click to place the view in the center of the drawing.

Click ESC to exit placing views.

8.
Modify the tolerance box by double left clicking on the text and updating.

9.
Delete the vertical line and the words Material and Finish.

10.
Select the **Note** tool.

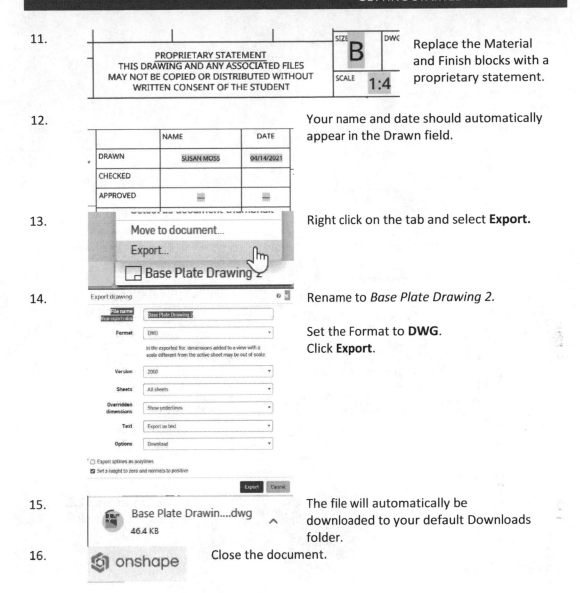

11. Replace the Material and Finish blocks with a proprietary statement.

12. Your name and date should automatically appear in the Drawn field.

13. Right click on the tab and select **Export.**

14. Rename to *Base Plate Drawing 2.*

Set the Format to **DWG.**
Click **Export.**

15. The file will automatically be downloaded to your default Downloads folder.

16. Close the document.

Scaling Views

Estimated Time: 10 minutes
Objectives:

- Modifying the scale of views

1.  Select the **Sample - Wind Meter - Copy** workspace.
This workspace was created in Chapter One.

2. Locate the tab for the **Base Plate Drawing 1**.

3. Highlight the top view.
Right click and select **View Properties**.

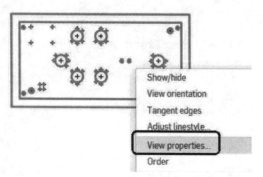

4. 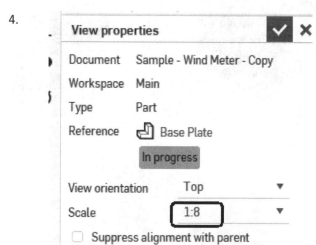 Change the scale to **1:8.**
Click the Green check.

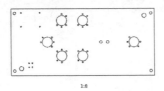

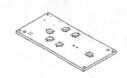

Only the top view updated.

1:8

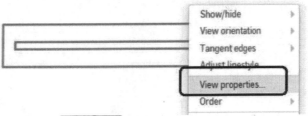

5. Click **Undo**.

6. Select the front view. Right click and select **View Properties**.

The front view is the parent view.

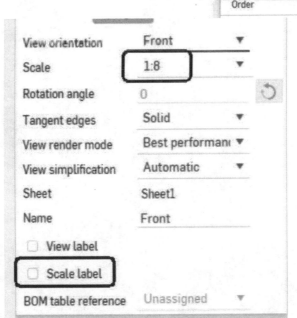

7. Change the Scale to **1:8.**

Disable **Scale label.**

Click **Green check.**

8.

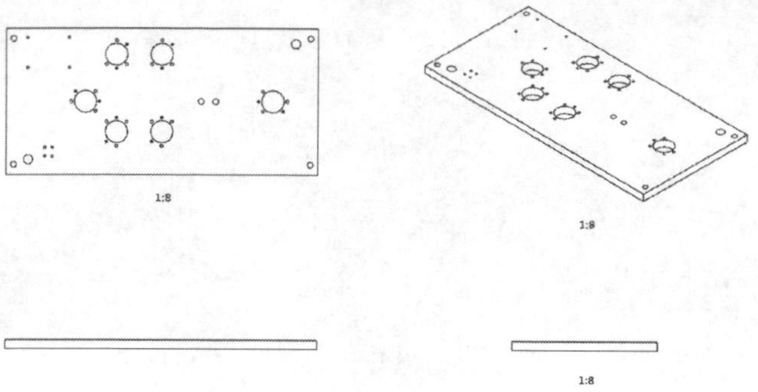

1:8

1:8

1:8

All the views updated.

9.   onshape   Close the document.

Place Drawing Views

Estimated Time: 25 minutes
Objectives:

- Place Drawing Views
- Change View to Show Hidden Lines
- Rename Tab
- Add Notes

1.

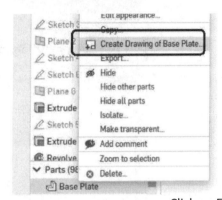

Select the **Sample - Wind Meter - Copy** workspace.
This workspace was created in Chapter One.

2.

Click on the **Part Studio 1** tab.

3.

Locate the **Base Plate** in the Parts panel of the browser.

Right click and select **Create drawing of Base Plate**.

4.

Click on Existing templates and select **Onshape**.

5.

Select the **ANSI_B_INCH** template.
Enable **No views**.
Click **OK.**

6.

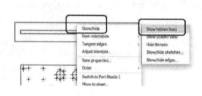

Select the **Top** view from the View Orientation drop-down list.
A preview will be displayed.

Change the View Scale to **1:8.**
Left click to place on the sheet.
Click to place the front view above the top view.
Click to ESC placing views.

7.

Select the front view.
Right click and select **Show/Hide→Show hidden lines**.

8.

3	2	1

Adjust the positions of the views.

9.

Select **Create projected view** from the toolbar.
Select the top view.

10.

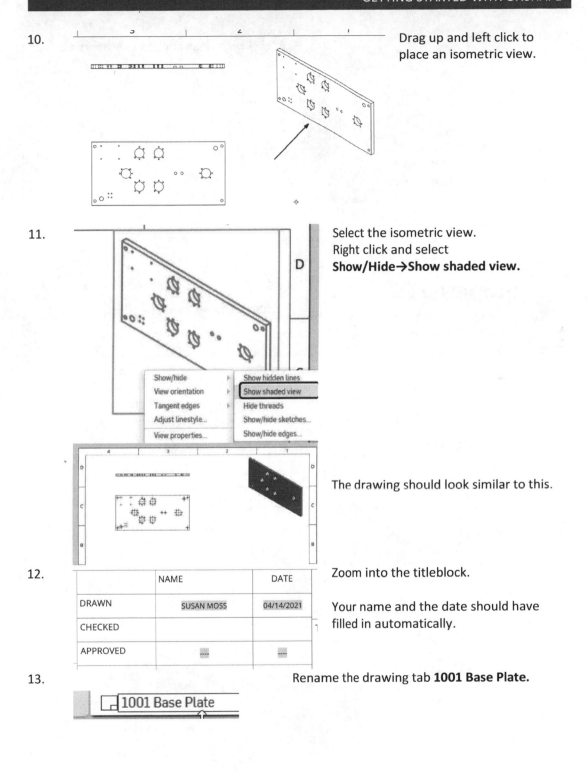

Drag up and left click to place an isometric view.

11.

Select the isometric view.
Right click and select
Show/Hide→Show shaded view.

The drawing should look similar to this.

12.

	NAME	DATE
DRAWN	SUSAN MOSS	04/14/2021
CHECKED		
APPROVED		

Zoom into the titleblock.

Your name and the date should have filled in automatically.

13.

1001 Base Plate

Rename the drawing tab **1001 Base Plate.**

14.

Right click on the tab and select **Properties**.

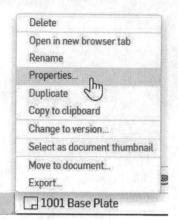

15.

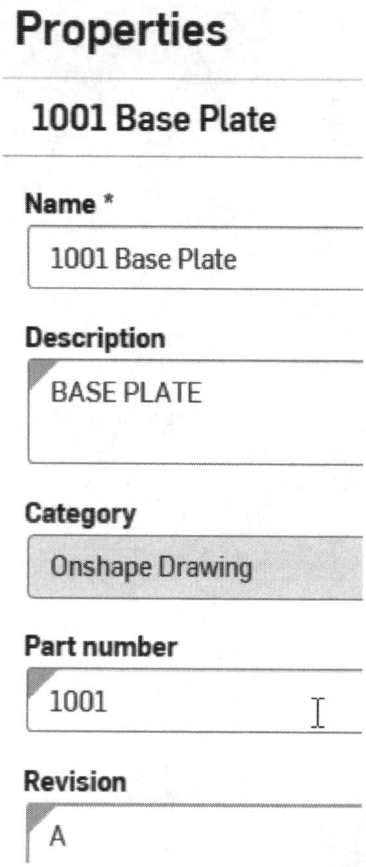

Properties

1001 Base Plate

Name *

1001 Base Plate

Description

BASE PLATE

Category

Onshape Drawing

Part number

1001

Revision

A

Type **BASE PLATE** in the Description field.

Type **1001** in the Part Number field.

Type **A** in the Revision field.

16.

State *

In progress

Title 1

BASE PLATE

Title 2

Title 3

Scroll down and type BASE PLATE in the Title 1 field.

Click **Save**.

17.

TITLE

BASE PLATE

SIZE
B

DWG NO.
1001

REV.
A

SCALE 1:8

WEIGHT

SHEET 1 of 1

The title block updates with the typed information.

If the title doesn't appear, click on the ---- and edit the text with the correct title.

Delete the unused title lines.

18.

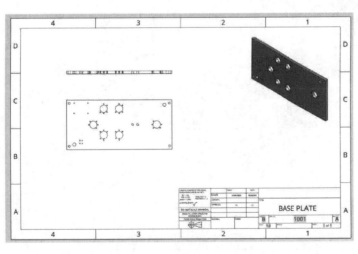

Close the document.

Adding Linear Dimensions

Estimated Time: 15 minutes

Objectives:

- Units
- Adding a linear dimension
- Change dimension precision
- Drawing Properties

1.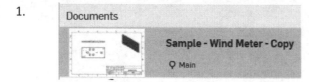

   Select the **Sample - Wind Meter - Copy** workspace.

   This workspace was created in Chapter One.

2. 1001 Base Plate

   Locate the tab for the **1001 Base Plate** Drawing.

   Remember you can drag and drop the tabs anywhere on the bar to make it easier to select and locate.

3.

   Select **Workspace Units** from the Documents menu.

4.

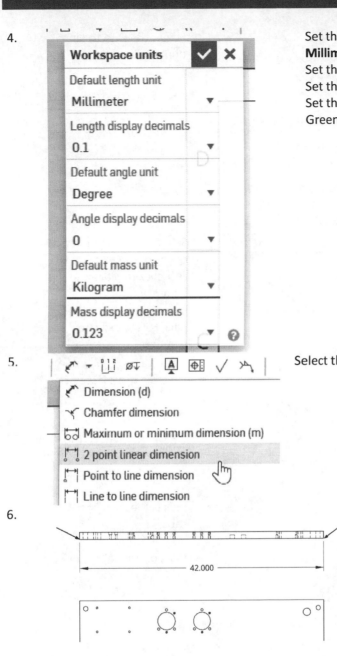

Set the Default length unit to **Millimeter.**
Set the Length display decimals to **0.1.**
Set the Angle display decimals to **0.**
Set the Default mass unit to **Kilogram.**
Green check.

5.

Dimension (d)
Chamfer dimension
Maximum or minimum dimension (m)
2 point linear dimension
Point to line dimension
Line to line dimension

Select the **2 point linear** dimension tool.

6.

42.000

Select the two end points of the bottom edge of the top view.
Pull the dimension down between the top and front views.
Left click to place.
Click ESC to exit the command.

7.

42.000

42.000   1.500   (in)
0
0.1
0.12
0.123
✓ 0.123 (Drawing)
0.1234

Double left click on the dimension's extension line.
Change the precision to one decimal place by using the drop-down box.
Left click on the sheet to exit.

8.

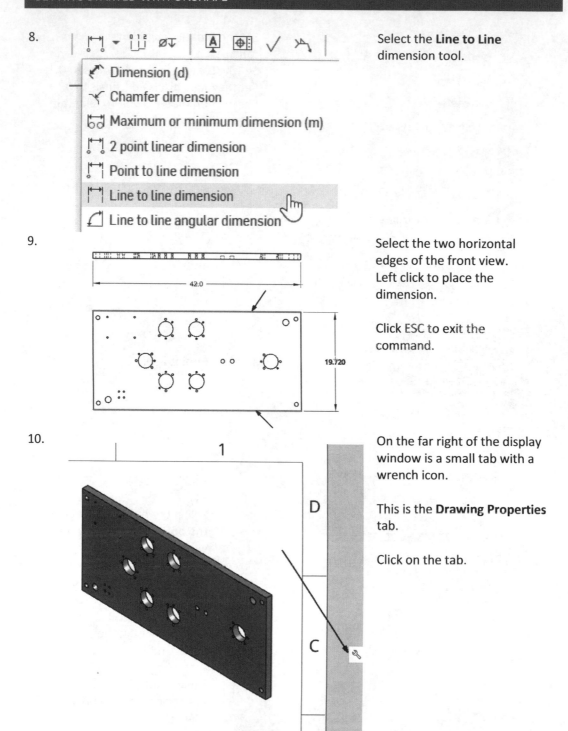

Dimension (d)

Chamfer dimension

Maximum or minimum dimension (m)

2 point linear dimension

Point to line dimension

Line to line dimension

Line to line angular dimension

Select the **Line to Line** dimension tool.

9.

42:0

19.720

Select the two horizontal edges of the front view. Left click to place the dimension.

Click ESC to exit the command.

10.

1

D

C

On the far right of the display window is a small tab with a wrench icon.

This is the **Drawing Properties** tab.

Click on the tab.

11.

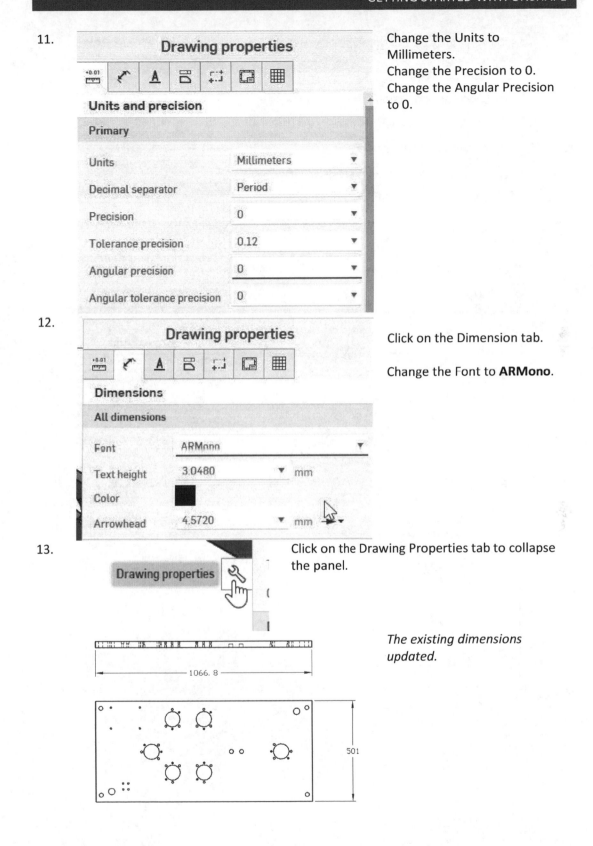

Change the Units to Millimeters.
Change the Precision to 0.
Change the Angular Precision to 0.

12.

Click on the Dimension tab.

Change the Font to **ARMono**.

13.

Click on the Drawing Properties tab to collapse the panel.

The existing dimensions updated.

14.

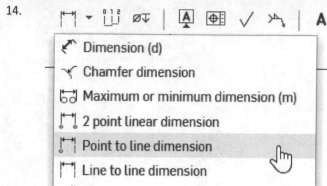

Select the **Point to Line** dimension tool.

15.

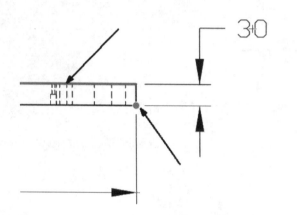

Select the top horizontal edge of the top view.
Select the lower right corner vertex.
Left click to place the dimension to the right of the view.
Click ESC to exit the command.

16.

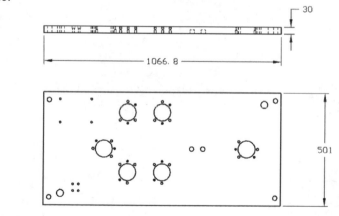

onshape

Close the document.

Adding a Sheet

Estimated Time: 10 minutes

Objectives:

- Add a sheet
- Add a view
- Change view scale

1.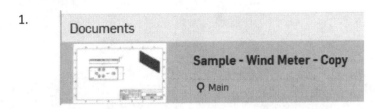

    Select the **Sample - Wind Meter - Copy** workspace. *This workspace was created in Chapter One.*

2. 1001 Base Plate

    Locate the tab for the **1001 Base Plate** Drawing.

3.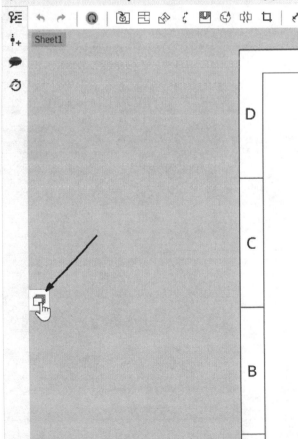

    On the left side of the display window is the Sheets tab.

    Click the **Sheets** tab or Click **Ctrl+S** on the keyboard.

4. 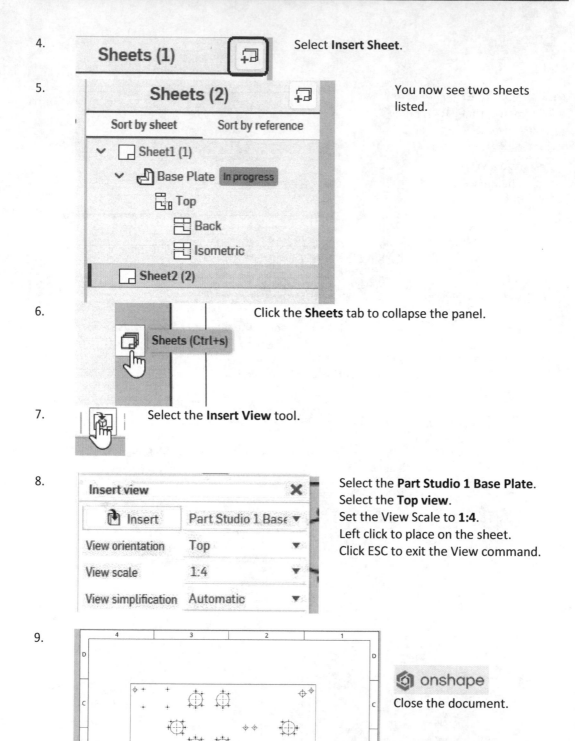 **Sheets (1)**   Select **Insert Sheet**.

5. **Sheets (2)**

You now see two sheets listed.

Sort by sheet   Sort by reference

∨ ☐ Sheet1 (1)

   ∨ ⬛ Base Plate In progress

      ⬛ Top

        ⬛ Back

        ⬛ Isometric

   ☐ Sheet2 (2)

6.   **Sheets (Ctrl+s)**   Click the **Sheets** tab to collapse the panel.

7.   Select the **Insert View** tool.

8. **Insert view**   ✕

   📋 Insert   Part Studio 1 Base ▼

   View orientation   Top   ▼

   View scale   1:4   ▼

   View simplification   Automatic   ▼

Select the **Part Studio 1 Base Plate**.
Select the **Top view**.
Set the View Scale to **1:4**.
Left click to place on the sheet.
Click ESC to exit the View command.

9.

🔷 onshape

Close the document.

Dimensioning Holes

Estimated Time: 20 minutes

Objectives:

- Add ordinate dimensions
- Add diameter dimensions
- Modify dimensions – adding a quantity

1. 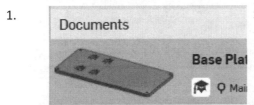 Select the **Base Plate** workspace.

   This was an exercise in Chapter 2.

2. 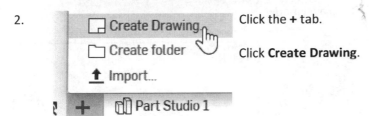 Click the **+** tab.

   Click **Create Drawing**.

3. 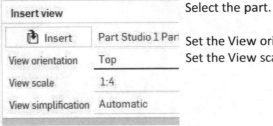 Click on the **ANSI_B_MM.dwt** template.

4. Enable **No views**.

   Click **OK**.

5.  Select the part.

   Set the View orientation to **Top**.
   Set the View scale to **1:4**.

6.

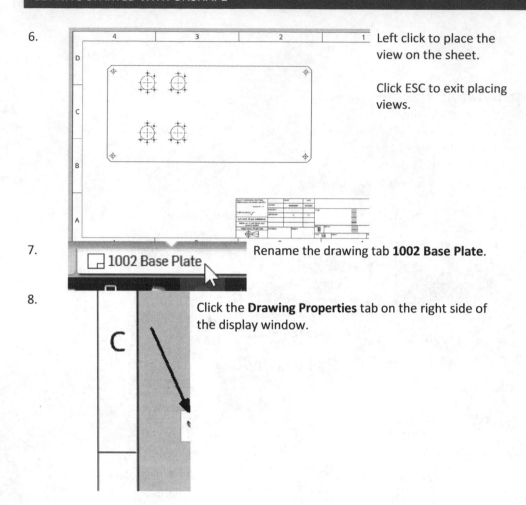

Left click to place the view on the sheet.

Click ESC to exit placing views.

7.   Rename the drawing tab **1002 Base Plate**.

8.   Click the **Drawing Properties** tab on the right side of the display window.

9.

Drawing properties

Units and precision

Primary

Units	Millimeters
Decimal separator	Period
Precision	0
Tolerance precision	0.12
Angular precision	0
Angular tolerance precision	0

Dual

☐ Show dual dimensions

☐ Show dual unit

Dimension location	Top
Hole callout location	Top
Units	Inches
Precision	0.123
Tolerance precision	0.123

Verify that the settings are correct.

Units should be Millimeters.

Precision should be set to 0.

Click the tab to collapse the panel.

10. Select the **Ordinate** dimension tool.

11.

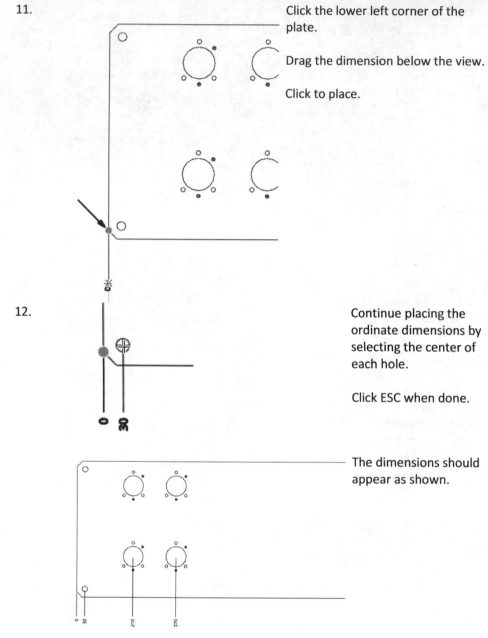

Click the lower left corner of the plate.

Drag the dimension below the view.

Click to place.

12.

Continue placing the ordinate dimensions by selecting the center of each hole.

Click ESC when done.

The dimensions should appear as shown.

13.   Select the **Ordinate** dimension tool.

14.

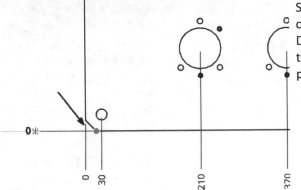

Select the bottom left corner of the plate. Drag the dimension to the left and left click to place.

15.

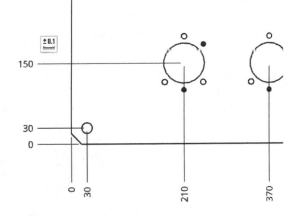

Continue to place vertical ordinate dimensions. Click **ESC** to exit the command.

16.   Select the **Hole Callout** tool.

The Hole Callout tool can only be used on features created as holes. If you extrude a circle using the Remove Option, it will not be considered a hole feature.

17.

Select the hole in the upper right corner.

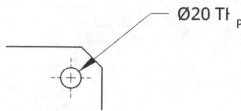

Ø20 Tŀ Place the hole dimension.

18.

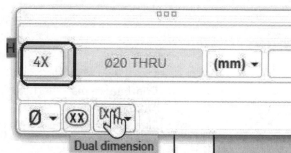

Double left click on the dimension.

Type **4X** in the prefix field.

Place a space after the 4X.

Left click in the window to close the edit panel.

19.

The hole callout updates.

4X Ø20 THRU

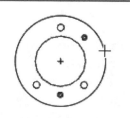

20. (Select the **Detail View** tool.

21.

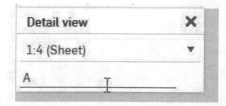

Draw a detail circle by selecting the center of the circle and then left picking to place the radius of the detail circle.

22.

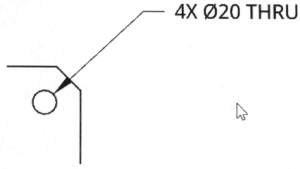

Set the Scale to **1:4.**

23.

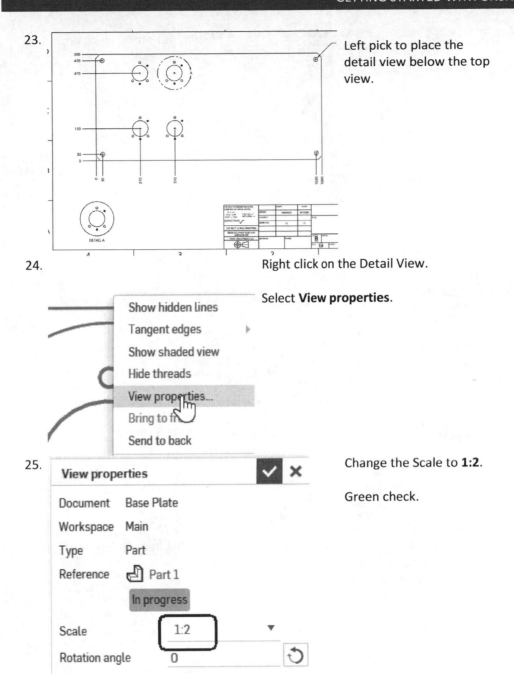

Left pick to place the detail view below the top view.

24.

Right click on the Detail View.

Show hidden lines

Tangent edges ▸

Show shaded view

Hide threads

View properties...

Bring to fr

Send to back

Select **View properties**.

25.

View properties ✓ ✕

Document   Base Plate

Workspace   Main

Type   Part

Reference   Part 1

In progress

Scale   1:2 ▾

Rotation angle   0 ↻

Change the Scale to **1:2**.

Green check.

26. 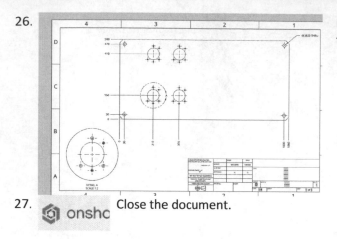 Reposition the views on the sheet.

27.  Close the document.

Adding Centerlines

Estimated Time: 10 minutes

Objectives:

- Add linear dimensions
- Add diameter dimensions
- Modify dimensions – adding a quantity
- Modify dimensions – adding symbols

1. Select the **Base Plate** workspace.

2. Open the tab for the **1002 Base Plate** Drawing.

3. Select the **2 pt Center line** tool.

4. Draw a center line from the upper corner hole to the lower corner hole.

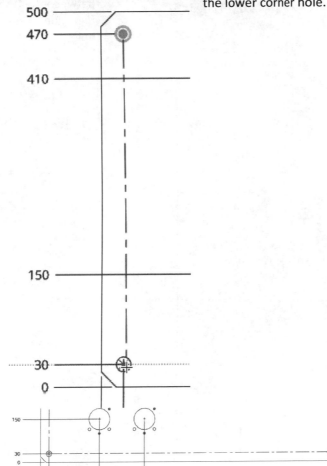

5. Draw a center line between the two bottom corner holes.

6.

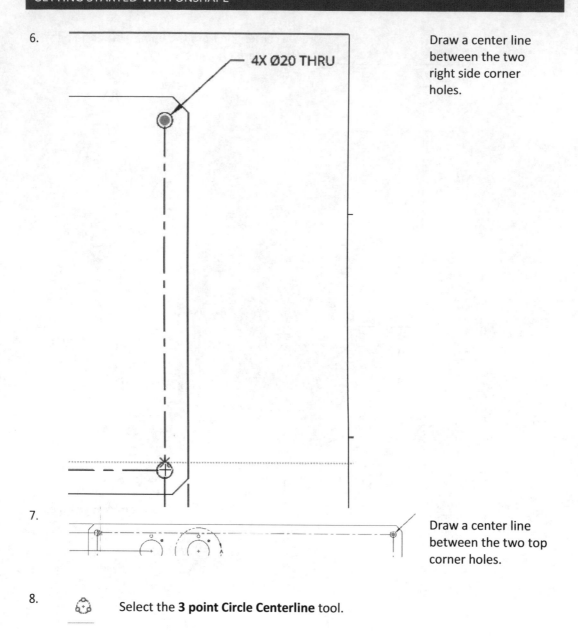

4X Ø20 THRU

Draw a center line between the two right side corner holes.

7.

Draw a center line between the two top corner holes.

8. Select the **3 point Circle Centerline** tool.

9.

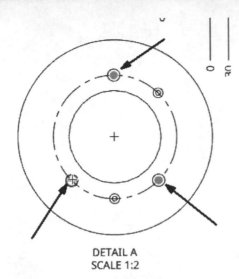

DETAIL A
SCALE 1:2

Select the three circles in the Detail View.

10.

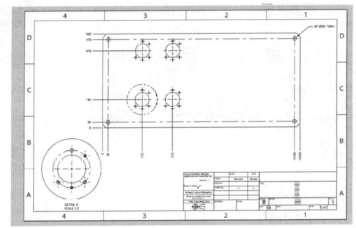

onshape

Close the document.

Creating a Hole Table

Estimated Time: 10 minutes
Objectives:
- Add Sheet
- Insert View
- Insert Sheet
- Insert Hole Table

1.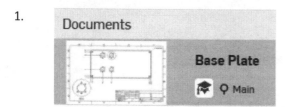
Select the **Base Plate** workspace.

2. Open the tab for the **1002 Base Plate** Drawing.

   □ 1002 Base Plate

3. Select the **Sheets** tab on the left side of the display window.

   Sheets (Ctrl+s)

4. **Sheets (1)**   Insert a Sheet.

5. Select **Insert View**.

6. Select the **Top** orientation.
Set the View Scale to **1:8**.
Left click to place the view on the sheet.
ESC to exit insert view.

   Insert view
   Insert Part Studio 1 Part ▼
   View orientation Top
   View scale 1:8
   View simplification Automatic

7. Insert a **Hole Table**.

8.

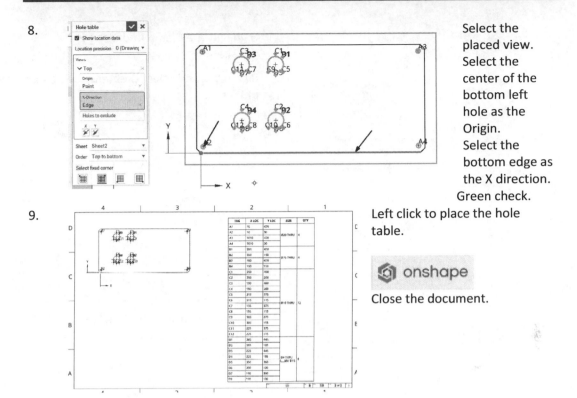

Select the placed view. Select the center of the bottom left hole as the Origin. Select the bottom edge as the X direction. Green check.

9.

Left click to place the hole table.

Close the document.

Creating a Section View

Estimated Time: 15 minutes

Objectives:

- Create drawing
- Assign Properties
- Add drawing view
- Change view display
- Create a section view

1.

Scooter
🎓 ♀ Main

Select the **Scooter** workspace.

2.

🗍 Clamp

Select the **Clamp** tab.

3.

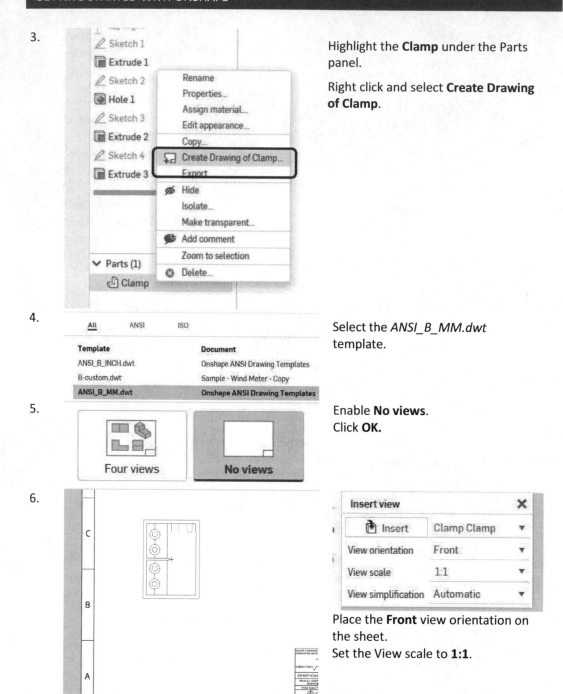

Highlight the **Clamp** under the Parts panel.

Right click and select **Create Drawing of Clamp**.

4.

Select the *ANSI_B_MM.dwt* template.

5.

Enable **No views**.
Click **OK.**

6.

Place the **Front** view orientation on the sheet.
Set the View scale to **1:1**.

7.

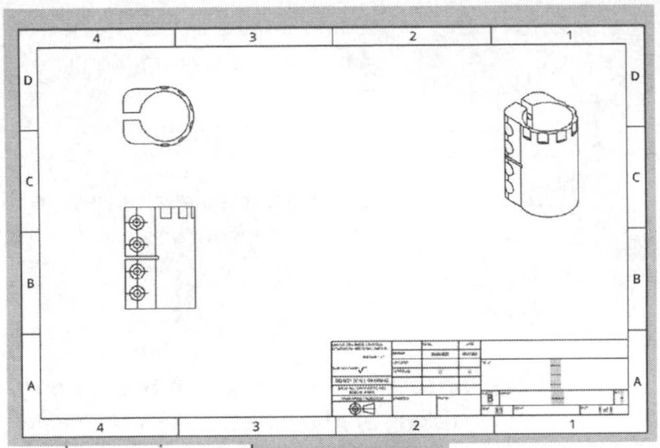

Place three views as shown:

- Front
- Top
- Isometric

8.

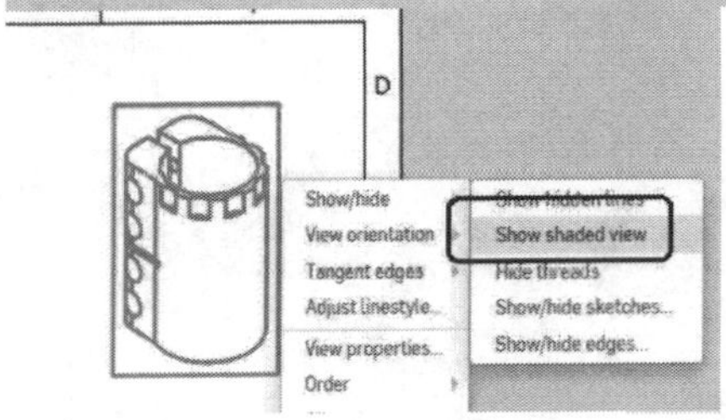

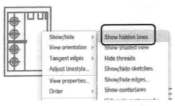

Select the **Isometric** view.
Right click and select **Show/hide→Show shaded view**.

9.

Select the **Front** view.
Right click and select **Show/hide→Show hidden lines**.

10.

Select the **Top** view.
Right click and select **Show/hide→Show hidden lines**.

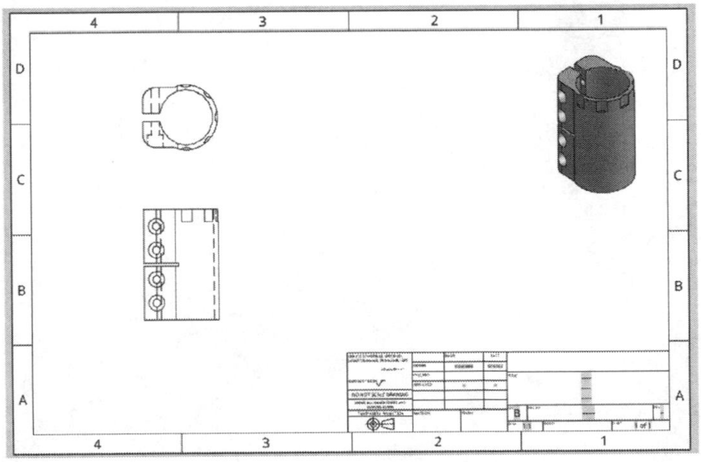

Your sheet should look similar to this.

11. Select the **Section view** tool.

12.

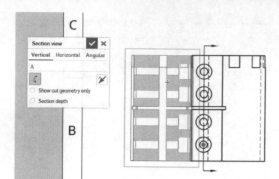

Select the center of the bottom hole to place the section line.

Enable Vertical.

Left click to place the view to the left of the front view.

13.

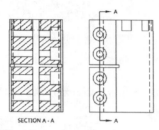

Place the section view to the left of the front view by left clicking on the sheet.

SECTION A - A

14.

Rename
Properties...
Duplicate
Copy to clipboard
Change to version...
Select as document thumbnail
Move to document...
Export...

Clamp | Clamp Drawing 1 | Fork

Right click on the tab.
Select **Properties.**

15.

Name *

Clamp

Description

CLAMP

Category

Onshape Drawing

Part number

1003

Revision

A

Change the Name to **Clamp**.
Type **CLAMP** in the Description field.
Type **1003** in the Part number field.
Type **A** in the Revision field.

16.

Title 1

CLAMP

Title 2

SCOOTER

Title 3

Add **CLAMP** to Title 1.
Add **SCOOTER** to Title 2.
Click **Save**.

TITLE		
	CLAMP	
	SCOOTER	
SIZE B	DWG NO. 1003	REV. A
SCALE 1:1	WEIGHT	SHEET 1 of 1

The title block should update.
If the titles don't update, you can click on the ---
text and edit.

17.

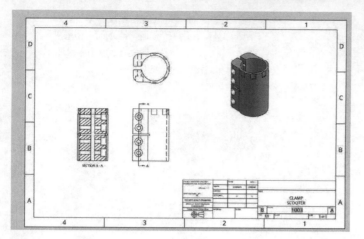

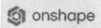

Close the document.

Extra: Dimension the drawing.

Creating an Auxiliary View

Estimated Time: 5 minutes
Objectives:

- Create drawing
- Add drawing view
- Create a projected view
- Create an auxiliary view

1.  Select the **Angled Plate** workspace.
This workspace was created in Chapter Two.

2.  Locate the Angled Plate in the Parts panel of the browser.
Right click and select **Create Drawing of Angled Plate.**

3. 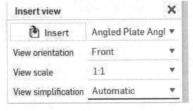 Select the *ANSI_B_MM* template under **Onshape templates**.
Enable **No views**.
Click **OK.**

4.  Select the **Angled Plate** from the drop-down list.
Select the **Front** view.
Set the View scale to **1:1**.
Left click to place on the sheet.
Click **ESC** to exit the insert view mode.

5.

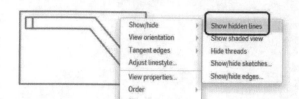

Right click on the front view and select **Show/hide→Show hidden lines**.

6.

Select the **Create Projected View** tool.

Select the front view.

7.

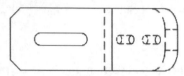

Drag up and left click to place a top view.

Notice that because the Front view was set to show hidden lines, the top view automatically used the same display properties.

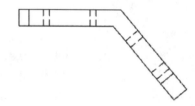

8.

Select the **Projected View** tool.

9.

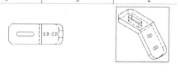

Select the front view.
Drag up and to the right, then left click to place an isometric view.

Move the isometric view over to the right to make room for an auxiliary view.

10.

Select the **Auxiliary View** tool.

11.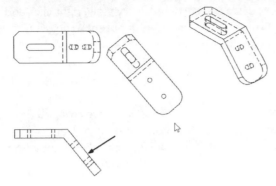

Select the angled top edge.
Place the auxiliary view to the right of the top view.

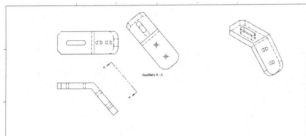

The sheet should have four views as shown.

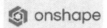

Close the document.

Creating a Cropped View

Estimated Time: 5 minutes
Objectives:

- Create Cropped View
- Change view alignment

1. Select the **Angled Plate** workspace.

Angled Plate

🎓 📍 Main

2. ☐ Angled Plate Drawing 1

Select the Drawing tab.

3. ⛶ Select the auxiliary view.
Select the **Crop View** tool.

4.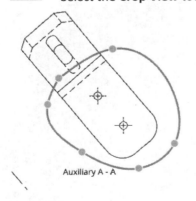

Draw a closed spline.

Profile must overlap one view only. ✕

3

If you get this error message, you will need to exit the command and move the auxiliary view away from the other views.

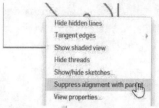

You can select the Auxiliary view, right click and select **Suppress alignment with parent** to adjust its position away from other views.

5.

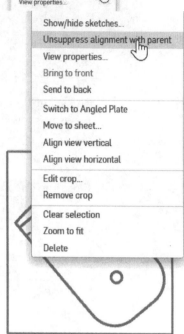

| Show/hide sketches... |
| Unsuppress alignment with parent |
| View properties... |
| Bring to front |
| Send to back |
| Switch to Angled Plate |
| Move to sheet... |
| Align view vertical |
| Align view horizontal |
| Edit crop... |
| Remove crop |
| Clear selection |
| Zoom to fit |
| Delete |

Once you have cropped the view, you can right click and select **Unsuppress alignment with parent** to re-align the view.

6.

Change the Isometric view so it is shaded and the hidden lines are removed.

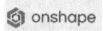

Close the document.

Dimensioning a Drawing

Estimated Time: 5 minutes

Objectives:

- Add dimensions
- Add Center Marks
- Change Drawing Properties

1. 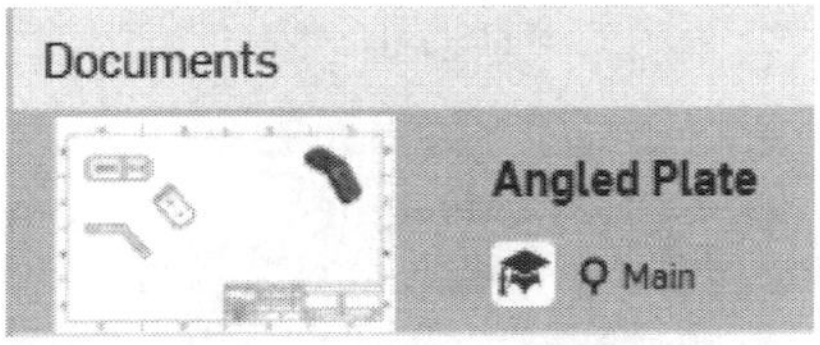   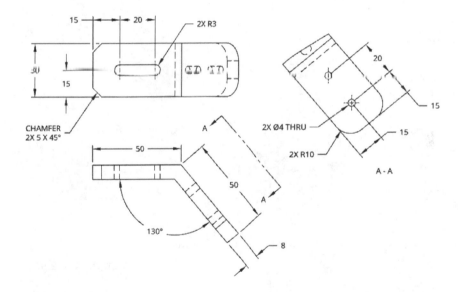   Select the **Angled Plate** workspace.

2. 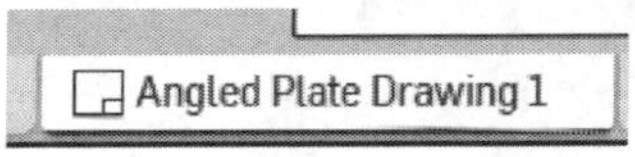   Select the Drawing tab.

3.   Add dimensions.

4.

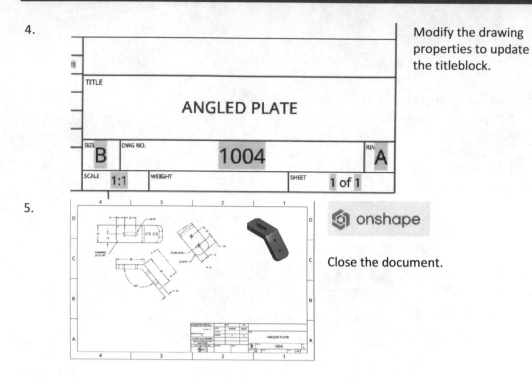

Modify the drawing properties to update the titleblock.

5.

onshape

Close the document.

Export to PDF

Estimated Time: 5 minutes
Objectives:
- Export a drawing tab to PDF

1.

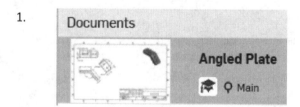

Open the **Angled Plate** document.

2.

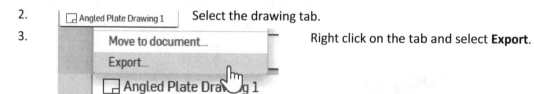

Select the drawing tab.

3.

Right click on the tab and select **Export**.

4.

File name		
View export rules	Angled Plate Drawing 1	
Format	PDF	▼
Overridden dimensions	Show underlines	▼
Text	Normal	▼
Color	Color	▼
Options	Download	▼

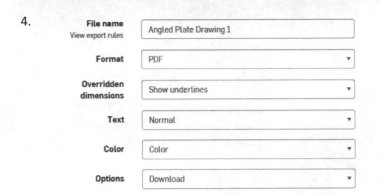

Set the Format to **PDF**.

Set the Options to **Download**.

Click **Export.**

The file will be downloaded to the Default Downloads folder.

Export   Cancel

5.

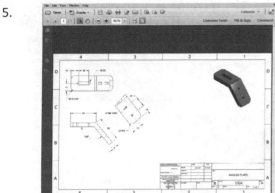

Open the file in Adobe Reader to check it.

6.

Close the document.

Extra: Create drawings for all the parts you have created and export to PDFs.

Assembly Drawing

Estimated Time: 5 minutes

Objectives:

- Create an assembly drawing
- Add item balloons

1.

   Open the **Scooter** assembly.

2.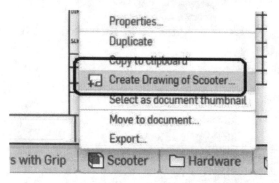

   Locate the **Scooter** Assembly in the tabs list.

   Right click and select **Create drawing of Scooter.**

3. All   ANSI   ISO

Template	Document
ANSI_B_MM.dwt	**Onshape ANSI Drawing Templates**
ANSI_B_INCH.dwt	Onshape ANSI Drawing Templates
B-custom.dwt	Sample - Wind Meter - Copy

   Show Onshape drawing templates

   Select *ANSI_B_MM.dwt*.
   Enable **No views**.

   Click **OK**.

   Four views

   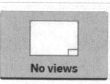
   No views

4.

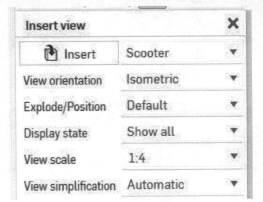

Select an **Isometric** view.

Set the View scale to **1:4**.
Left click to place on the sheet.

5.

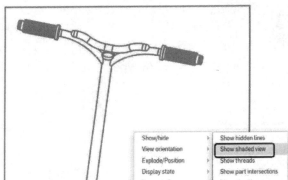

Right click on the view and select **Show/hide→Show shaded view**.

6. Select the **Item balloon** tool.

7.

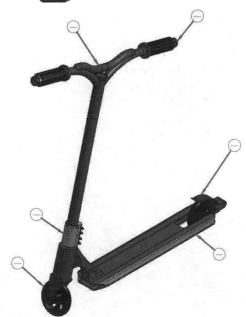

Add item balloons.
Green check.

8. Close the document.

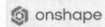

Parts Lists

Estimated Time: 5 minutes
Objectives:

- Create a parts list

1.

   Open the **Scooter** assembly.

2.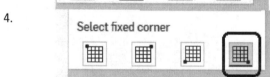

   Select the **Insert BOM Table** tool.

3.

   Select the Structured – Top level BOM type.

4. Select the bottom right corner to place the table.

5.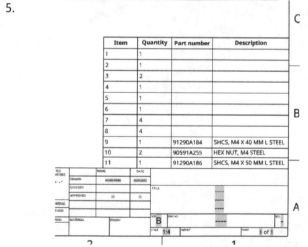

   Left click to place on the sheet above the title block.

Item	Quantity	Part number	Description
1	1		
2	1		
3	2		
4	1		
5	1		
6	1		
7	4		
8	4		
9	1	91290A184	SHCS, M4 X 40 MM L STEEL
10	2	90591A255	HEX NUT, M4 STEEL
11	1	91290A186	SHCS, M4 X 50 MM L STEEL

6.

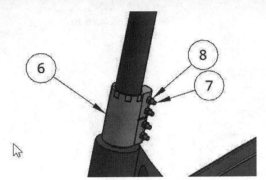

Add the Item Balloons to call out the bolt and nut on the clamp.

Onshape still doesn't offer the ability to gather or stack item balloons. This may change in the future.

We are missing information on several of the parts.

Item	Quantity	Part number	Description
1	1		
2	1		
3	2		
4	1		
5	1		
6	1		
7	4		
8	4		
9	1	91290A184	SHCS, M4 X 40 MM L STEEL
10	2	90591A255	HEX NUT, M4 STEEL
11	1	91290A186	SHCS, M4 X 50 MM L STEEL

NAME	DATE
SURNEMOSE	12/06/2022

7.

Select the **Scooter** assembly tab.

8.

Select the **BOM** tab on the right side of the display window.

9.

When you click in a cell, the corresponding part will highlight in the display window to assist you in filling in the blanks.

If you are using hardware that is a public document, you will need to create a copy in order to change the properties.

10.

Return to the drawing tab.

11.

BOM table properties...				
Copy		umber	Description	
Move to ▸				
Bring to front				
Send to back				
Switch to BOM : Scooter				
Move to sheet...				
Clear selection		A258	SCHS M5 X 40 MM L	
Zoom to fit		A300	METRIC HEX NUT M5 X 0.8MM	
Delete		A184	SHCS, M4 X 40 MM L STEEL	
		A255	HEX NUT, M4 STEEL	
11	1	91290A186	SHCS, M4 X 50 MM L STEEL	

The BOM did not update.

12.

Click **Update**.

Item	Quantity	Part number	Description
1	1	2001	DECK
2	1	2002	BRAKE
3	2	2003	WHEEL
4	1	2004	FORK
5	1	2005	HANDLEBARS
6	2	2006	GRIPS
7	1	2007	CLAMP
8	4	91290A258	SCHS M5 X 40 MM L
9	4	92497A300	METRIC HEX NUT M5 X 0.8MM
10	1	91290A184	SHCS, M4 X 40 MM L STEEL
11	2	90591A255	HEX NUT, M4 STEEL
12	1	91290A186	SHCS, M4 X 50 MM L STEEL

The BOM should update/refresh.

13.

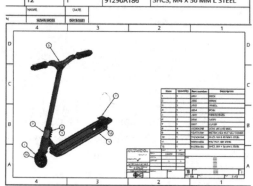

⊙ onshape

Close the workspace.

Create an Exploded View

Estimated Time: 30 minutes
Objectives:

- Create an Exploded View
- Named View

1.  Open the **Scooter** assembly.

2.  Select the **Scooter** assembly tab.

3.   Select the **Exploded Assembly** tab located on the right of the display window.

4. 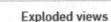 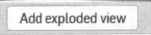 Click on **Add exploded view**.

5. Highlight **Handlebars with Grips** in the browser.

    Left click on the handlebars in the display window to activate the triad.

6. 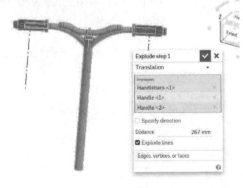 Use the triad to move the sub-assembly up and away from the deck.

    Verify that both grips and handlebars are selected.

    Note you can adjust the distance in the dialog.
    Enable **Explode lines**.
    Green check.

    Left click in the display window to release the selection.

7.

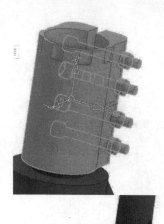

Highlight the Clamp and the bolts and nuts that are inserted in the clamp.

8.

Use the triad to move the clamp and hardware between the fork and the handlebars.

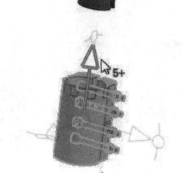

Explode step 2	✓	✗
Translation	▼	
Instances		
Clamp <1>		✗
91290A258 <1>		✗
92497A300 <4>		✗
☐ Specify direction		
Distance	125 mm	
☐ Explode lines		❓

Disable the **Explode lines**.
Green check.
Left click in the display window to release the selection.

9.

Select the four nuts placed on the clamp.

10.

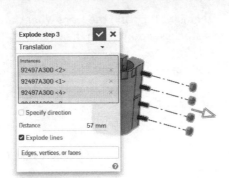

Use the triad to move the nuts to the right of the clamp.

Enable the **Explode lines**.
Green check.
Left click in the display window to release the selection.

11.

Select the four bolts placed on the clamp.

12.

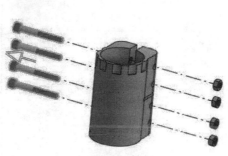

Use the triad to move the nuts to the left of the clamp.

13.

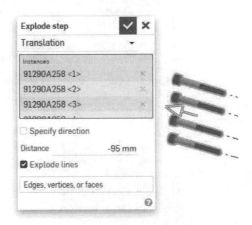

Enable the **Explode lines**.
Green check.
Left click in the display window to release the selection.

14.

Highlight the fork, the front wheel, and the fasteners mounted on the wheel.

You can use a window to select the hardware.

15.

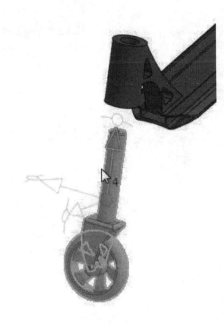

Use the triad to move the selection below the deck.

Enable the **Explode lines**.
Green check.
Left click in the display window to release the selection.

16.

Select the cap screw.

17.

Use the triad to move the selection to the left.

Explode step 6	✓ ✗
Translation	▼

Instances

91290A184 <2>   ×

☐ Specify direction

Distance   151 mm

☑ Explode lines

Edges, vertices, or faces

Enable the **Explode lines**.
Green check
Left click in the display window to release the selection.

18.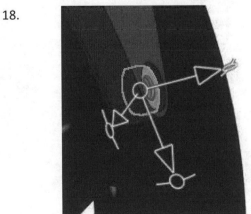

Select the nut mounted on the fork.

I orbited the view to locate and select the nut.

19.

Use the triad to move the selection to the right.

20.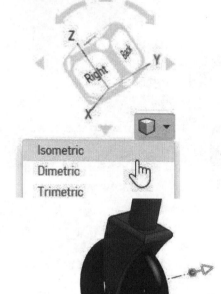

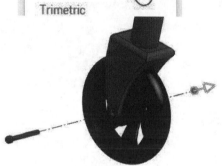

Return the view to the default isometric to verify that the nut is located properly in the exploded view.

21.

Adjust the nut distance if needed.

Explode step 7	✓	✗

Translation ▼

Instances
90591A255 <1>   ✗

☐ Specify direction

Distance   80 mm

☑ Explode lines

Edges, vertices, or faces

Enable the **Explode lines**.
Green check.
Left click in the display window to release the selection.

22.

Select the front wheel.

23.

Position the wheel away from the fork.

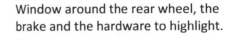

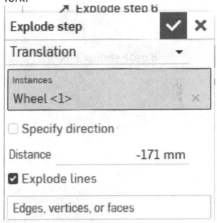

Enable the **Explode lines**.
Green check.
Left click in the display window to release the selection.

24.

Window around the rear wheel, the brake and the hardware to highlight.

25.

Move the selection above the deck.

Disable the **Explode lines**.

Green check.

Left click in the display window to release the selection.

26.

Select the brake.

27.

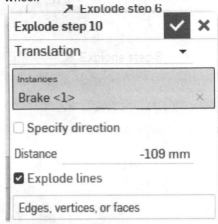

Move the brake to the right of the wheel.

Enable the **Explode lines**.
Green check.
Left click in the display window to release the selection.

28.

Select the cap screw in the rear wheel.

29.

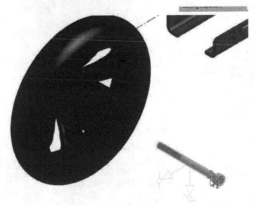

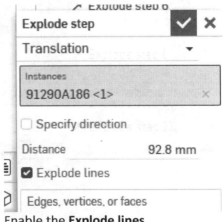

Move the cap screw away from the rear wheel.

Enable the **Explode lines**.
Green check.
Left click in the display window to release the selection.

30.

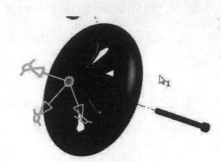

Orbit the view so you can select the nut on the rear wheel.

31.

Move the nut away from the wheel.

32.

Switch back to an Isometric view.

33.

Adjust the position of the nut by modifying the Distance value as needed.

Enable the **Explode lines**.
Green check.
Left click in the display window to release the selection.

34.

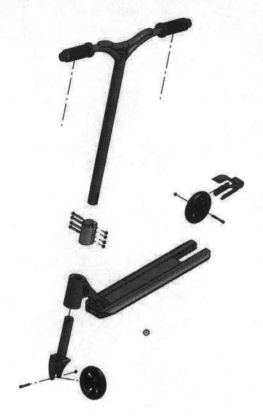

Your exploded view should look similar to this.

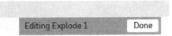

Click **Done** to saved the exploded view.

35.

To see the exploded view, highlight **Explode 1**.

Right click and select **Activate Explode 1.**

36.

Click on the drop-down next to the display cube.
Select **Named views...**

37.

Type **Iso exploded**.
Click **+**.

The named view is listed.

38.   Close the document.

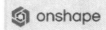

Place an Exploded View on a Sheet

Estimated Time: 15 minutes

Objectives:

- Add Sheet
- Place an Exploded View
- Add Item Balloons
- Modify Item Balloons
- Insert BOM

1. 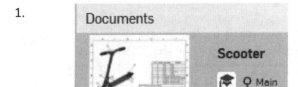 Open the **Scooter** assembly workspace.

2. Select the **Scooter** drawing tab.

   Scooter Drawing 1

3. Select the Sheets tab on the left of the display window.

4. **Sheets (1)** Select the **Add Sheet** tool.

   Sort by sheet   Sort by reference

   ∨ ☐ Sheet1 (1)

   　∨ ▣ Scooter In progress

   　　 Isometric

   　　 Scooter

5. Select the **Insert View** tool.

6.

Insert view ✕

📥 Insert | Scooter ▼

View orientation | Iso exploded ▼

Explode/Position | Explode 1 ▼

Display state | Show all ▼

View scale | 1:8 ▼

View simplification | Automatic ▼

Set the View orientation to **Iso exploded.**

Set Explode to **Explode 1.**

Set the View scale to **1:8**.

Left click to place the view on the sheet.

7.

Show/hide | ▶ | Show hidden lines
View orientation | ▶ | Show shaded view
Explode/Position | ▶ | Show threads

Right click on the exploded view.
Select **Show/hide→Show shaded view**.

8. Add **Item Balloons**.

9.

Edit...
Copy
Add leader
Remove leaders
Move to ▶
Clear selection
Zoom to fit
Delete

8

Select Item Balloon for the clamp bolts.

Right click and select **Edit**.

10.

Callout ✓ ✕

11

3.0480 ▼ Circle | 2 Characters ▼

None
Circle
Diamond
Double Circle
Flag
Hexagon
Octagon
Rectangle
Split Circle
Square Circle
Triangle
Underline

Select **Split Circle**.

11
4

Type **4** in the lower text box.

Green check.

11.

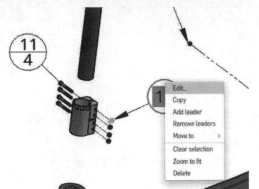

Select Item Balloon for the clamp nuts.

Right click and select **Edit**.

12.

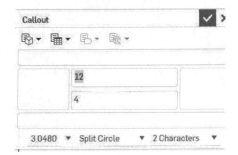

Select **Split Circle**.

Type **4** in the lower text box.

Green check.

13.   Insert the BOM.

14.

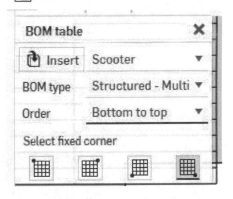

Set the BOM type to **Structured-Multi-level**.
This is an indented BOM.
Set the Order to **Bottom to top**.
Set the insert point to the **lower right**.

15.

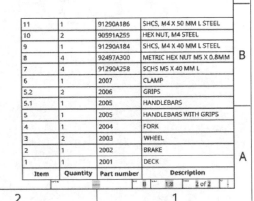

Click to place in the lower right corner of the drawing.

Item	Quantity	Part number	Description	
11	1	91290A186	SHCS, M4 X 50 MM L STEEL	
10	2	90591A255	HEX NUT, M4 STEEL	
9	1	91290A184	SHCS, M4 X 40 MM L STEEL	
8	4	92497A300	METRIC HEX NUT M5 X 0.8MM	B
7	4	91290A258	SCHS M5 X 40 MM L	
6	1	2007	CLAMP	
5.2	2	2006	GRIPS	
5.1	1	2005	HANDLEBARS	
5	1	2005	HANDLEBARS WITH GRIPS	
4	1	2004	FORK	
3	2	2003	WHEEL	
2	1	2002	BRAKE	A
1	1	2001	DECK	

16.

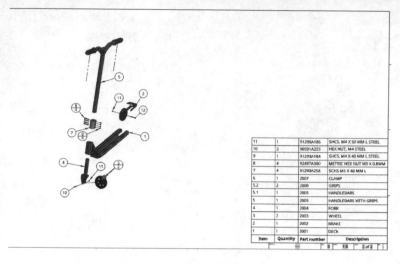

Item	Quantity	Part number	Description
11	1	91290A186	SHCS, M4 X 50 MM L STEEL
10	2	90591A255	HEX NUT, M4 STEEL
9	1	91290A184	SHCS, M4 X 40 MM L STEEL
8	4	92497A300	METRIC HEX NUT M5 X 0.8MM
7	4	91290A258	SCHS M5 X 40 MM L
6	1	2007	CLAMP
5.2	2	2006	GRIPS
5.1	1	2005	HANDLEBARS
5	1	2005	HANDLEBARS WITH GRIPS
4	1	2004	FORK
3	2	2003	WHEEL
2	1	2002	BRAKE
1	1	2001	DECK
Item	Quantity	Part number	Description

Close the document.

Export a Parts List

Estimated Time: 5 minutes
Objectives:

- Export a parts list

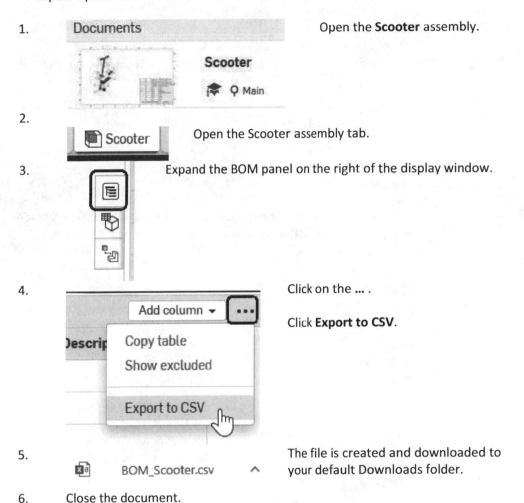

1.   Open the **Scooter** assembly.

2.   Open the Scooter assembly tab.

3.   Expand the BOM panel on the right of the display window.

4.   Click on the **...** .

                                                          Click **Export to CSV**.

5.   The file is created and downloaded to your default Downloads folder.

6.   Close the document.

Create a Sheet Metal Drawing with Flat Pattern

Estimated Time: 10 minutes

Objectives:

- Insert a Flat Pattern View
- Insert Views
- Change the view display

1.

Documents

Scooter

🎓 ○ Main

Open the **Scooter** assembly.

2.

📗 Brake

Open the **Brake** tab.

3.

Click on the **Sheet Metal and Flat View** tab on the right side of the display window.

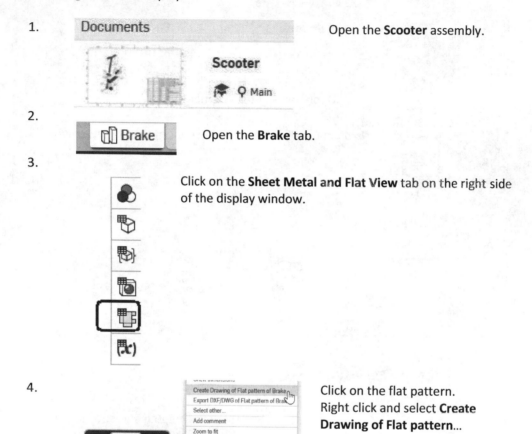

4.

Create Drawing of Flat pattern of Brake

Export DXF/DWG of Flat pattern of Brake

Select other...

Add comment

Zoom to fit

Zoom to selection

View normal to

Delete Sheet metal model 1

Click on the flat pattern.
Right click and select **Create Drawing of Flat pattern**...

5.

All   ANSI

Select the **ANSI_B_MM.dwt** template.

Template

ANSI_B_INCH.dwt

ANSI_B_MM.dwt

B-custom.dwt

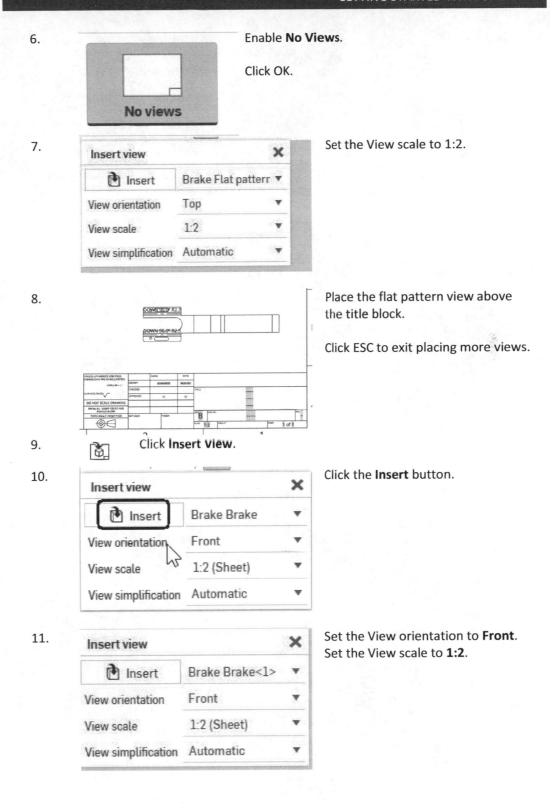

6.　Enable **No Views**.

Click OK.

7.　Set the View scale to 1:2.

8.　Place the flat pattern view above the title block.

Click ESC to exit placing more views.

9.　Click **Insert View**.

10.　Click the **Insert** button.

11.　Set the View orientation to **Front**.
Set the View scale to **1:2**.

12.

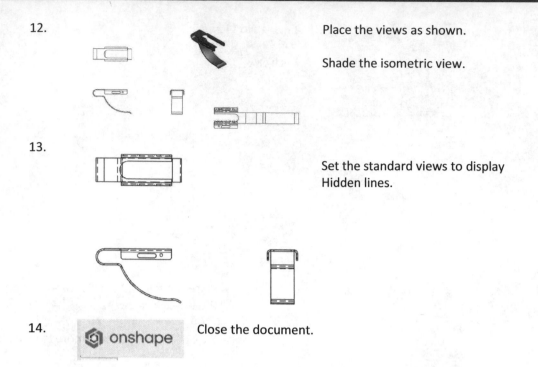

Place the views as shown.

Shade the isometric view.

13.

Set the standard views to display Hidden lines.

14.

Close the document.

Extra: Add dimensions.

Update the Drawing Properties so the title block displays the correct description, title, and revision.

Add center marks.

Add hole callouts.

Notes:

Chapter 9: Stop Base Project

The Stop Base

Estimated Time: 45 minutes
Objectives:

- Set Work Units for a Document
- Rename Tabs
- Extrude
- Rib
- Draft
- Create Hole
- Using Construction Lines
- Extrude
- Fillet

1.

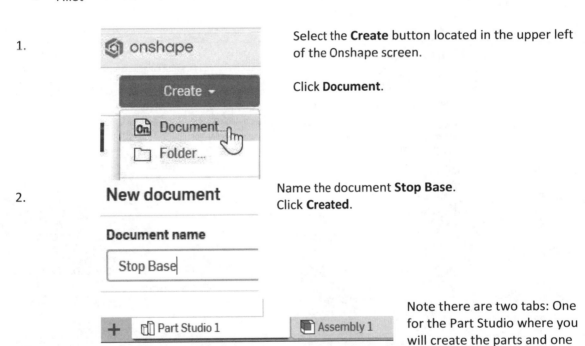

Select the **Create** button located in the upper left of the Onshape screen.

Click **Document**.

2.

Name the document **Stop Base**.
Click **Created**.

Note there are two tabs: One for the Part Studio where you will create the parts and one for Assembly.

3.

Select the drop down under the Documents menu.
Select **Workspace Units**.

4.

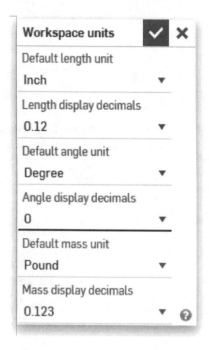

Set the Default length unit to **Inch**.
Set the Length display decimals to **0.12**.
Set the Angle display decimals to **0**.
Set the Default mass unit to **Pound**.
Click the **Green check**.

5.

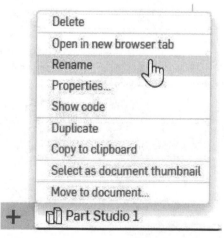

Select the Part Studio 1 tab. Right click and select **Rename.**

6. 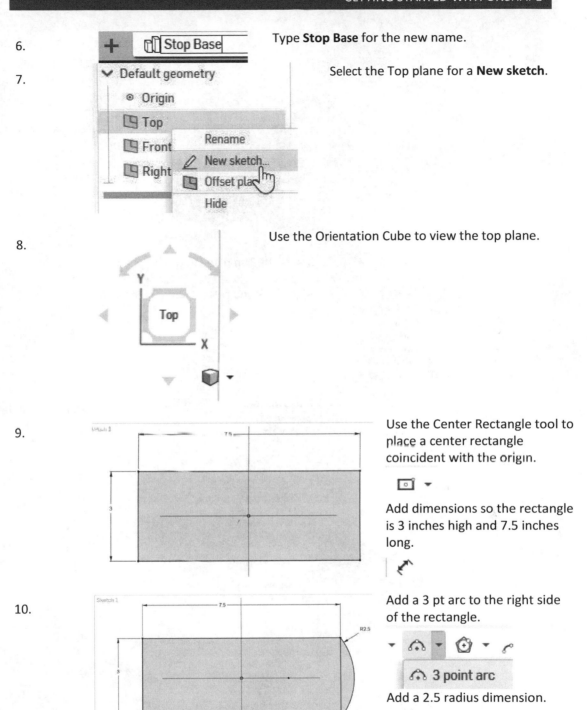 Type **Stop Base** for the new name.

7. Select the Top plane for a **New sketch**.

8. Use the Orientation Cube to view the top plane.

9. Use the Center Rectangle tool to place a center rectangle coincident with the origin.

Add dimensions so the rectangle is 3 inches high and 7.5 inches long.

10. Add a 3 pt arc to the right side of the rectangle.

3 point arc

Add a 2.5 radius dimension.

11.

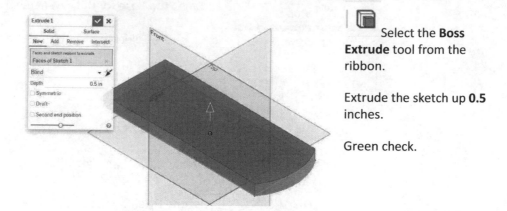 Select the **Boss Extrude** tool from the ribbon.

Extrude the sketch up **0.5** inches.

Green check.

12.

Select the Top plane in the browser.

Right click and select **Hide all planes**.

13.

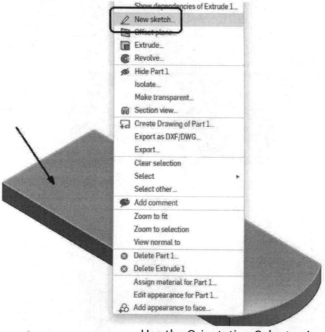

Select the top face of the part for a new sketch.

14.

Use the Orientation Cube to view the top plane.

15. 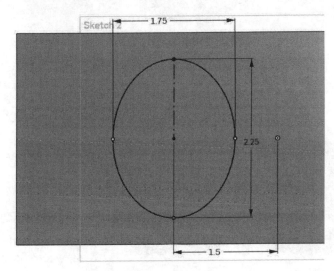 Select the **Ellipse** tool.

Center point circle [C]

3 point circle

Ellipse

16.

Add four points – one at each quadrant to assist in adding dimensions.
Add a vertical reference line to control the orientation of the ellipse, if needed.

Dimension the ellipse so it is located 1.5 inches to the left of the origin.

Add a horizontal constraint between the center of the ellipse and the origin.

The ellipse should be 1.75 inches wide and 2.25 inches high.

17. Select the **Boss Extrude** tool from the ribbon.

18.

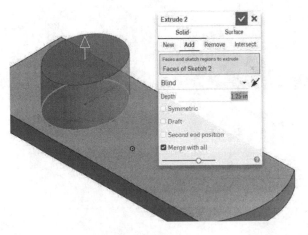

Enable **Add**.

Set the Depth to **1.25** inches.

Enable **Merge with all**.

Green check.

19.

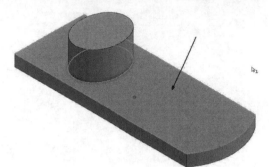

Select the top face of the first extrude.

 Sketch

Select **Sketch** from the ribbon.

20.

Use the Orientation Cube to view the top plane.

21.

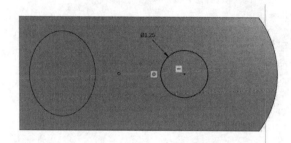

Draw a center point circle.

Add a **1.25** diameter dimension.

Add a concentric constraint between the circle and the arc.

22.

Select the **Boss Extrude** tool from the ribbon.

23.

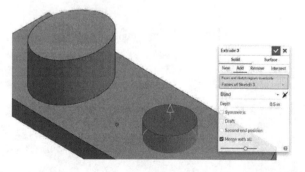

Enable **Add**.

Set the Depth to **0.5** inches.

Enable **Merge with all**.

Green check.

24.

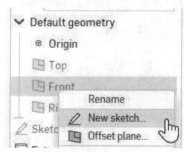

Select the **Front** plane for a new sketch.

25. Switch to a **Front** view.

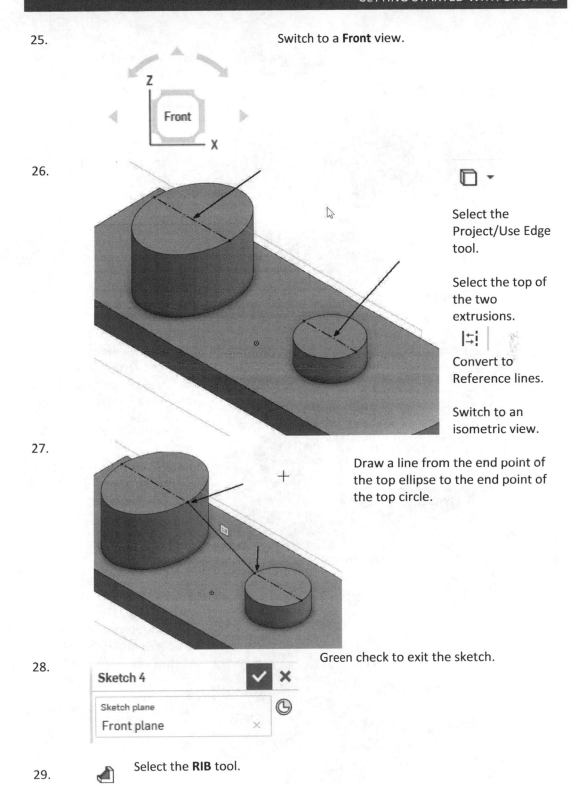

26. Select the Project/Use Edge tool.

Select the top of the two extrusions.

Convert to Reference lines.

Switch to an isometric view.

27. Draw a line from the end point of the top ellipse to the end point of the top circle.

Green check to exit the sketch.

28.

Sketch 4	✓	✗
Sketch plane		
Front plane	✕	

29. Select the **RIB** tool.

30.

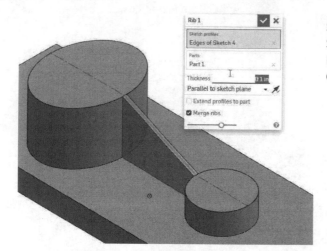

Select the slanted line.
Set the Thickness to **0.1**.
Enable **Merge ribs**.
Green check.

31. Select the **Draft Face** tool.

32.

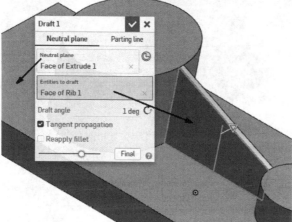

Select the top face of the extrude as the Neutral plane.
Select the vertical face of the rib as the entities to draft.
Set the Draft angle to **1°**.
Green check.

33. Select the **Draft Face** tool.

34.

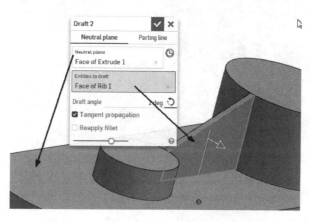

Select the top face of the extrude as the Neutral plane.
Select the vertical face of the rib as the entities to draft.
Set the Draft angle to **1°**.

Green check.

35.

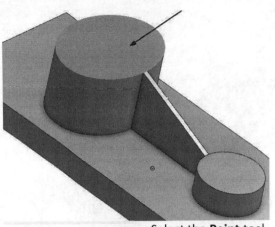

Place a new sketch on the top of the elliptical extrude.

36. Select the **Point** tool.

37.

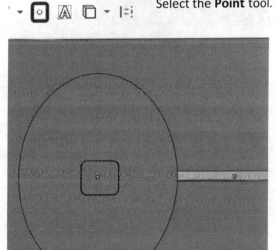

Place the point so it is concentric to the ellipse.

Green check to exit the sketch.

38. Select the **Hole** tool from the Features ribbon.

39.

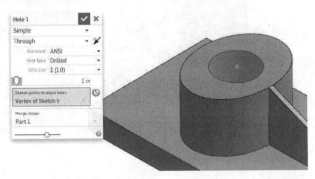

Set the Hole type to **Simple**.
Set the Distance to **Through**.
Select **ANSI**, **Drilled**, and set the size to **1** inch.

Select the point.
Green check.

40.

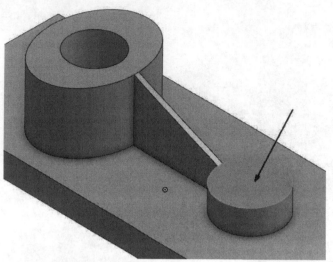

Place a new sketch on the top of the cylindrical extrude.

41.

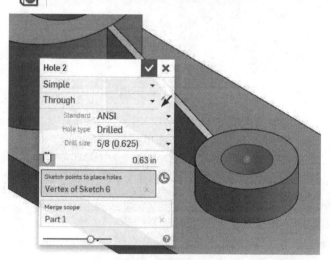

Place the point so it is concentric to the cylinder.

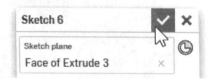

Green check to exit the sketch.

42. 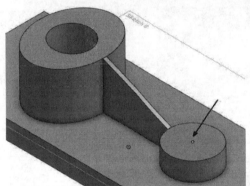 Select the **Hole** tool from the Features ribbon.

43.

Set the Hole type to **Simple**.
Set the Distance to **Through**.
Select **ANSI**, **Drilled**, and set the size to **5/8** inch.

Select the point.
Green check.

44.

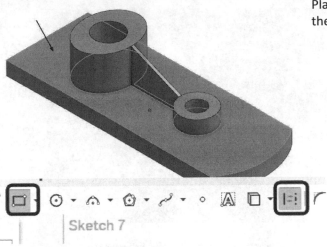

Place a new sketch on the top of the base extrude.

45.

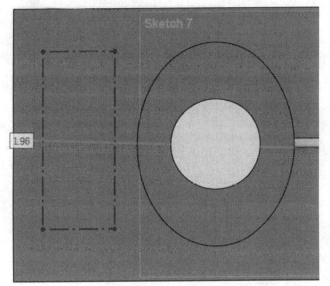

Select the **Corner Rectangle** sketch tool and set it to **Construction**.

46.

Draw a rectangle to the left of the ellipse.

47.

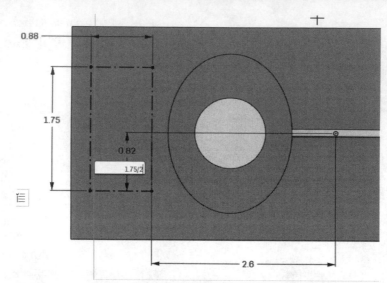

Add dimensions to constrain the rectangle.

Set the Height to **1.75**.
Set the Width to **0.875**.
Set the x-distance to **2.6** to the origin.
Set the y-distance to **1.75/2** to the origin.

Select the green check to exit the sketch.

48. Select the **Hole** tool from the Features ribbon.

49.

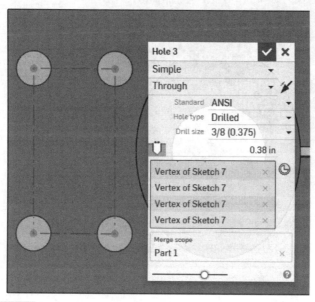

Set the Hole type to **Simple**.
Set the Distance to **Through**.
Select **ANSI**, **Drilled**, and set the size to **3/8** inch.

Select the vertex corner points of the rectangle to place four holes. Green check.

50. Select the **Fillet** tool from the Features ribbon.

51.

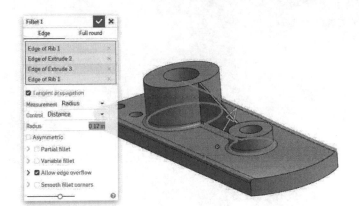

Select the two long sides of the base extrude. Select the base of the rib and the two extrudes. Set the Fillet radius to **0.12**.

Green check.

52.

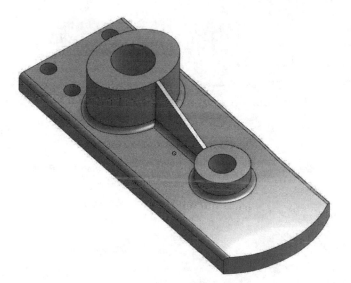

Close the document.

Using Standard Content

Estimated Time: 5 minutes

Objectives:

- Insert Part
- Insert Standard Content

1.

    Open the **Stop Base** workspace.

2.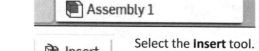

    Open the **Assembly1** tab.

3. Insert   Select the **Insert** tool.

4.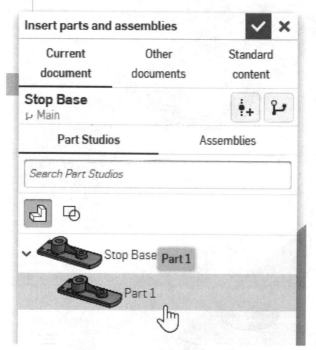

    Click on the Part1 of the Stop Base to insert into the Assembly.

    Green check.

5. Insert   Select the **Insert** tool.

6.

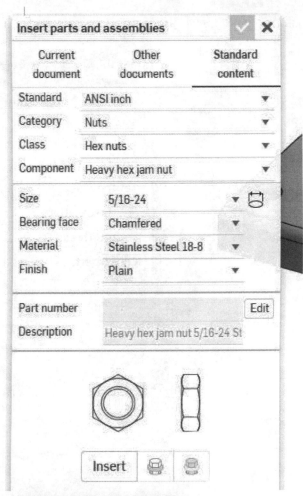

Click on **Standard content**.

Select Standard **ANSI Inch**.

Set the Category to **Nuts**.

Set the Class to **Hex Nuts**.

Set the Size to **5/16-24.**

Set the Material to **Stainless Steel 18-8.**

Click **Insert.**

7.

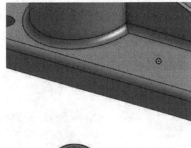

Left click to place in the display window.

8.

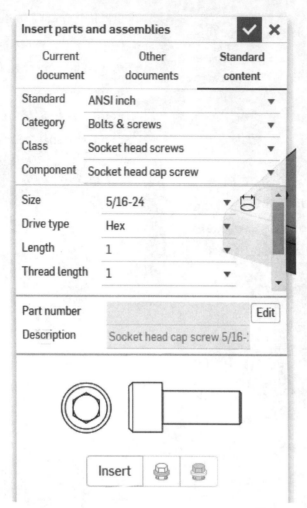

Click on Standard content.

Select Standard **ANSI Inch**.

Set the Category to **Bolts & Screws.**

Set the Class to **Socket head screws.**

Set the Component to **Socket head cap screw.**

Set the Size to **5/16-24.**

Set the Length to **1**.

Set the Thread Length to **1**.

Click **Insert.**

9.

Left click to place in the display window.

Green check to close the Insert dialog.

10.

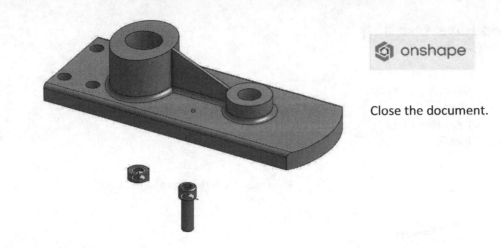

Close the document.

Stop Base Assembly
Estimated Time: 10 minutes
Objectives:

- Change Properties
- Rename Tab
- Fastened Mate
- Replicate

1.

   Open the **Stop Base** workspace.

2.

   Rename the Assembly tab.

3.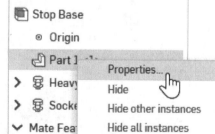

   Type **Stop Base**.

4. | Instances (3) |
   | Stop Base |
   | Origin |
   | Part 1 |
   | > Heavy |
   | > Socke |
   | ∨ Mate Fea |

   | Properties... |
   | Hide |
   | Hide other instances |
   | Hide all instances |
   | Isolate... |

   Highlight **Part 1** in the browser.

   Right click and select **Properties**.

5.

Part 1

Name *

STOP BASE

Description

STOP BASE

Category

Onshape Part

Part number

1001

Revision

A

Type **STOP BASE** in the Name Field.

Type **STOP BASE** in the Description Field.

Type **1001** in the Part Number field.

Type **A** in the Revision Field.

Part 1

Product line

Title 1

STOP BASE

Type **STOP BASE** in the Title 1 field.

Click **Save**.

6.

Instances (3)

Stop Base

⊙ Origin

STOP BASE <1>

> Heavy hex jam nut 5/1...

> Socket head cap scre...

✔ Mate Features (0)

The name updates in the browser.

∨ Parts (1)

STOP BASE

+   Stop Base

If you click the part tab, you will see the name updated in the part studio tab as well.

7.

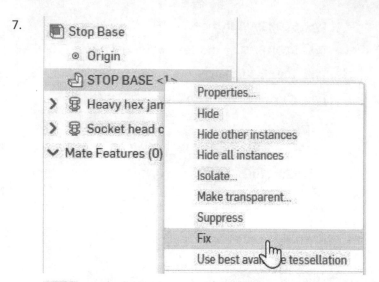

Right click on the Stop Base part.

Select **Fix.**

8. Select the **Fastened** mate.

9.

Select the underside of the cap screw head.

Select the top edge of the 3/8 hole.

Green check.

Close the Fastened panel.

10. Select the **Replicate** tool.

11.

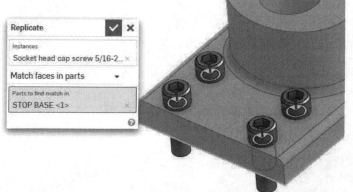

Select the cap screw.

Left click in the Faces to find match in field.

Select the top face.

Green check.

12.

Four screws are placed.

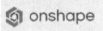

Close the workspace.

Creating a Named View

Estimated Time: 10 minutes

Objectives:

- Saving a view

1.

Open the **Stop Base** workspace.

2.

Activate the **Stop Base** Assembly tab.

3.

Orient the Orientation Cube as shown with Left and Front isometric.

4.

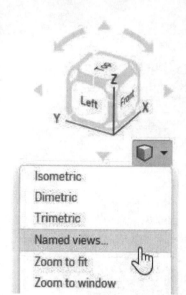

Left click on the Display Cube.

Select **Named views**.

5.

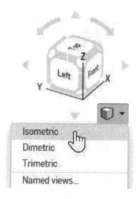

Type **rear iso**.

Left click on the add view icon.

Close the panel.

6.

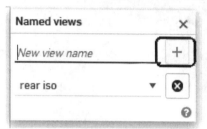

Left click on the Display Cube.

Select **Isometric**.

The display updates.

7.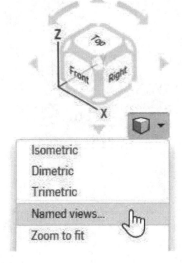

Left click on the Display Cube.

Select **Named views**.

8.

Select the **rear iso** view from the drop-down list.

Note how the display updates.

9. Close the document.

Extra:

Create an exploded view of the Stop Base Assy.

Create a drawing for the Stop Base part and for the Assembly.

Design a matching plate for the assembly and mount the nuts below the plate.

Chapter 10: Pulley Project

The Pulley

Estimated Time: 20 minutes

Objectives:

- Set Units for a Document
- Rename Tabs
- Create Hole
- Using Construction Lines
- Extrude
- Fillet

1.

    Select the **Create** button located in the upper left of the Onshape screen.

    Click **Document**.

2. **New document**

    Document name

    Pulley

    Name the document **Pulley**.
    Click **Create**.

    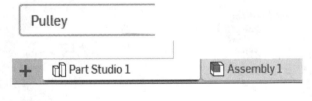

    Note there are two tabs: One for the Part Studio where you will create the parts and one for Assembly.

3.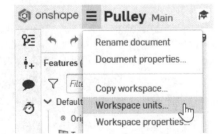

    Select the drop down under the Documents menu.
    Select **Workspace Units**.

4.

Set the Default length unit to **Millimeter**.
Set the Length display decimals to **0.1**.
Set the Angle display decimals to **0**.

Set the Default mass unit to **Kilogram**.
Click the **Green check**.

5.

Select the Part Studio 1 tab. Right click and select **Rename.**

6.

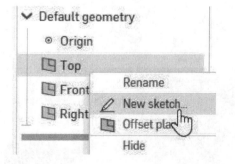

Type **Pulley Base** for the new name.

7.

Select the Top plane for a **New sketch**.

8. Use the Orientation Cube to view the top plane.

9.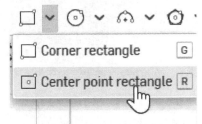

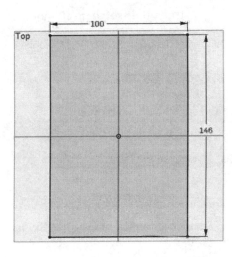

Use the Center Rectangle tool to place a center rectangle coincident with the origin.

Add dimensions so the rectangle is **146 mm** x **100 mm**.

10.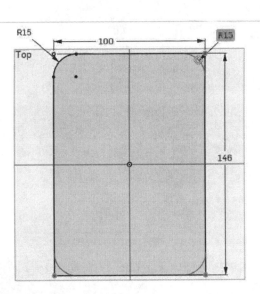

Add a **R15** fillet to the four corners of the rectangle.

Hint: Be sure to click the left mouse button to accept each fillet.

11.

Select the **Boss Extrude** tool from the ribbon.

Extrude the sketch up **10 mm**.

Green check.

12.

Select the Top plane in the browser.

Right click and select **Hide all planes**.

13.

Select the top face of the part for a new sketch.

14.

Switch the view orientation to **Top**.

15.

Select the **Center Rectangle** sketch tool and set it to **Construction**.

16.

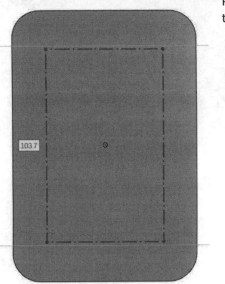

Place a center rectangle on the top face of the extrude.

17.

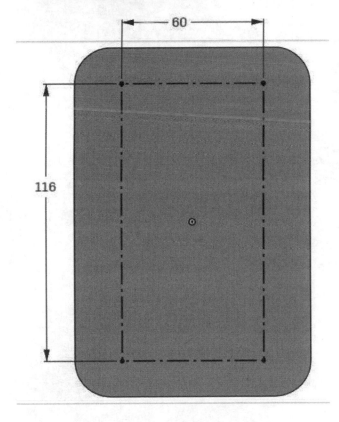

Add dimensions to constrain the rectangle.

Set the Height to **116 mm**.
Set the Width to **60 mm**.
Set the center coincident to the origin.

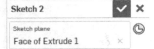

Select the green check to exit the sketch.

18.   Select the **Hole** tool from the Features ribbon.

19.

Set the Hole type to **Simple**.
Set the Distance to
Through.
Select **ISO**, **Drilled**, and set
the size to **10 mm**.

Select the vertex corner
points of the rectangle to
place four holes.
Green check.

20.

✓ Parts (1)

Pulley Base

Rename the Part in the browser **Pulley Base**.

21.

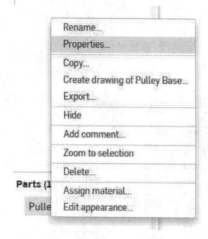

Right click on the Pulley Base part.
Select **Properties**.

22.

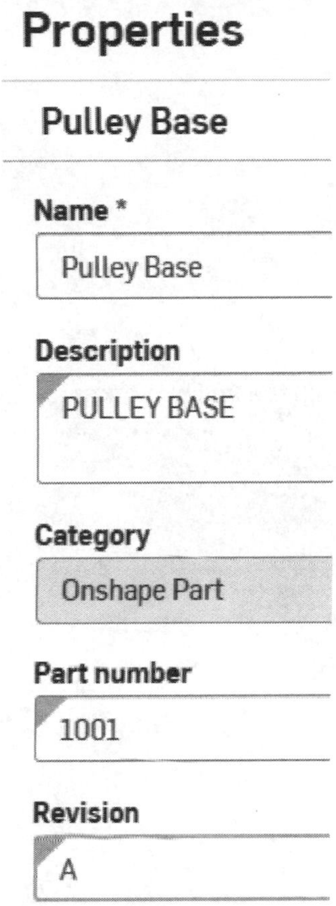

Properties

Pulley Base

Name *

Pulley Base

Description

PULLEY BASE

Category

Onshape Part

Part number

1001

Revision

A

Type **PULLEY BASE** in the Description field.
Type **1001** in the Part number field.
Type **A** in the Revision field.

Click **Save**.

23.

onshape

Close the workspace.

The Pulley Bracket

Estimated Time: 30 minutes

Objectives:

- Rename Tabs
- Create Hole
- Using Construction Lines
- Extrude
- Fillet

1.    Open the **Pulley** Workspace.

2. 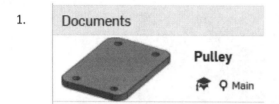   Select the + tab and left click on **Create Part Studio.**

3.    Right click on the Part Studio tab and select **Rename.**

4.   Type **Pulley Bracket**.

5.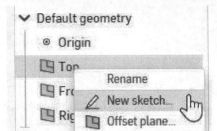

Select the **Top** plane for a **New sketch**.

6.

Switch to a **TOP** view orientation.

7.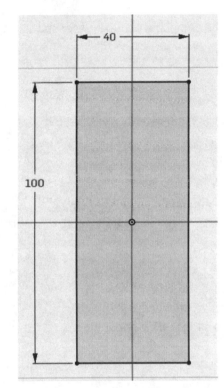

Use the Center Rectangle tool to place a center rectangle coincident with the origin.

Add dimensions so the rectangle is **40 mm** x **100 mm**.

8.

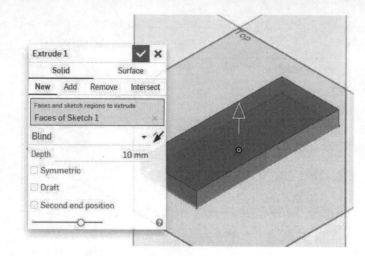

 Select the **Boss Extrude** tool from the ribbon.

Extrude the sketch up **10 mm**.

Green check.

9.

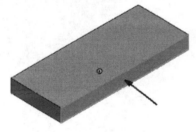

Right click on the Top plane in the browser and select **Hide all planes**.

10.

Select the right face of the extrude. Right click and select **New sketch**.

11.

Switch to a **Right** view orientation.

12.

Project/Use the top edge of the extrude into the sketch.

Draw a circle directly above the origin.

Draw two lines from the endpoints of the projected horizontal line to be tangent to the circle.

13.

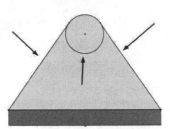

14.

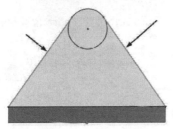

Select the two angled lines.
Add an **equal** constraint.

=

15.

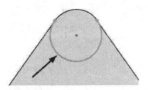

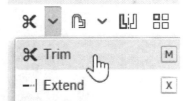

Use the **Trim** tool to trim the lower part of the circle.

Add dimensions.

16.

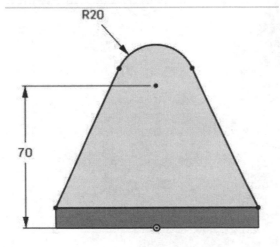

Set the radius of the arc to **R20** mm.

Set the vertical distance from the origin to the arc center to **70 mm**.

Notice that the sketch is now fully constrained – the color changes to black.

17.

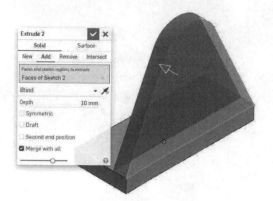

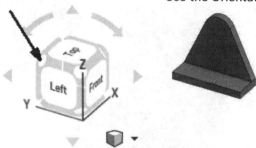

Select the **Boss Extrude** tool from the ribbon.

Enable **Add**.

Extrude the sketch **10 mm** into the part. *Use the arrows to flip the direction.*

Enable **Merge with all.**

Green check.

18.

Use the Orientation Cube to orient the part as shown.

19.

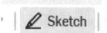

Select **Sketch** on the Features ribbon.
Select the front face of the part.

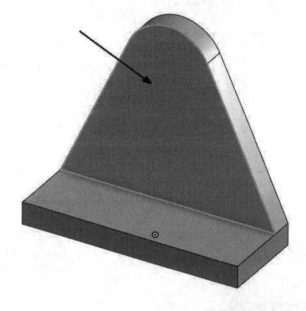

20.

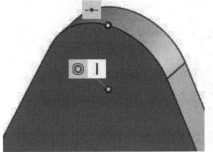

Place a point concentric to the arc.

Green check to exit the sketch.

21.      Select the **Hole** tool from the Features ribbon.

22.

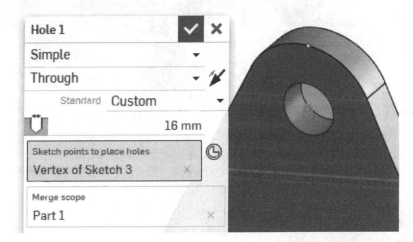

Set the Hole type to **Simple**. Set the Distance to **Through**. Select **ISO**, **Drilled**, and set the size to **16 mm**.

Select the concentric point. Green check.

23.   Select the **Fillet** tool from the Features ribbon.

24.

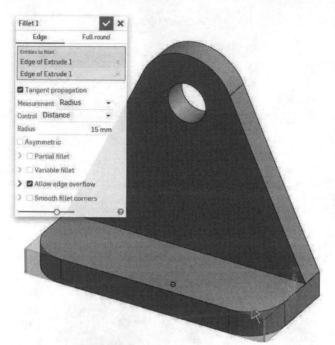

Select the front two corners and set the radius to **15** mm.

Green check.

25.

Select the top face for a new sketch.

26. 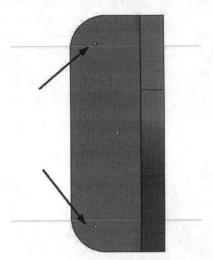 Place two points – **but do not make them concentric to the fillets.**

27. 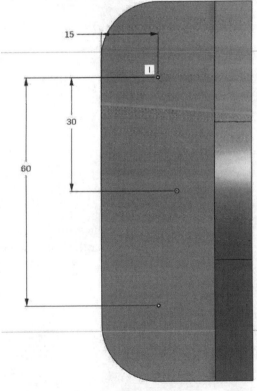 Add dimensions.

Add a **15 mm** from the left edge to the points.

Add a **60 mm** vertical dimension between the points.

Add a **30 mm** dimension between the top point and origin.

Add a vertical constraint between the two points.

Green check to exit the sketch.

28. Select the **Hole** tool from the Features ribbon.

29.

Hole 2 ✓ ✕
Simple   ▾
Through   ▾ ✐
   Standard   ISO   ▾
   Hole type Drilled   ▾
   Drill size 10   ▾
   🖰   10 mm
Sketch points to place holes 🕐
Vertex of Sketch 4   ✕
Vertex of Sketch 4   ✕
Merge scope
Part 1   ✕
   ───────○──────────   ❓

Set the Hole type to **Simple**.
Set the Distance to **Through**.
Select **ISO**, **Drilled**, and set
the size to **10 mm**.

Select the two points.
Green check.

30.

Select the back face of the bracket for a new sketch.

31.

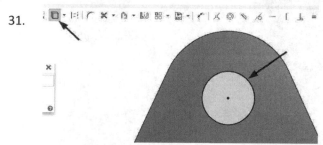

Use the **Project/Use** tool to project the hole edge into the sketch.

32.

Draw a circle.
Make it concentric to the hole.
Add a tangent constraint between the circle and the arc.

33.

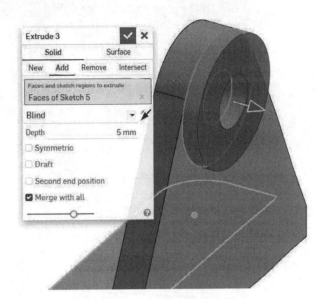

Select **Extrude**.

Enable **Add**.
Extrude **5 mm**.
Enable **Merge with all**.
Green check.

34.

Parts (1)

Pulley Bracket

Rename the Part in the browser **Pulley Bracket**.

35.

Right click on the Pulley Bracket part.
Select **Properties**.

36.

Properties

Pulley Bracket

Name *

Pulley Bracket

Description

PULLEY BRACKET

Category

Onshape Part

Part number

1002

Revision

A|

Type **PULLEY BRACKET** in the Description field.
Type **1002** in the Part number field.
Type **A** in the Revision field.

Click **Save**.

37.   🔷 onshape   Close the workspace.

The Pulley Wheel

Estimated Time: 40 minutes
Objectives:
- Rename Tabs
- Create Hole
- Using Construction Lines
- Extrude
- Fillet

1.

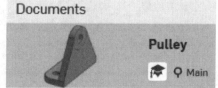

Open the **Pulley** Workspace.

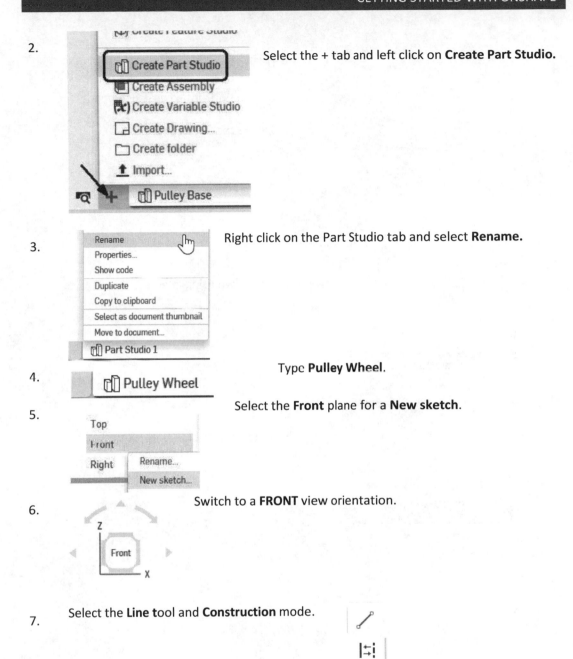

2. Select the + tab and left click on **Create Part Studio.**

3. Right click on the Part Studio tab and select **Rename.**

4. Type **Pulley Wheel**.

5. Select the **Front** plane for a **New sketch**.

6. Switch to a **FRONT** view orientation.

7. Select the **Line tool** and **Construction** mode.

8.

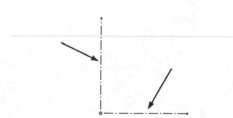

Draw a vertical construction line.

Draw a horizontal construction line.

Escape the line command.

9.

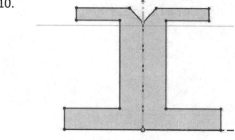

Draw the shape shown using lines.

10.

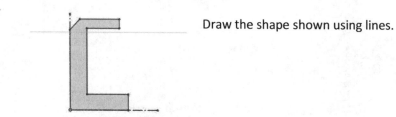

Use the **Mirror** tool to mirror the sketch about the vertical construction line.
Select the vertical construction line.
Select the lines to mirror.
Right click and select **Escape mirror**.

11.

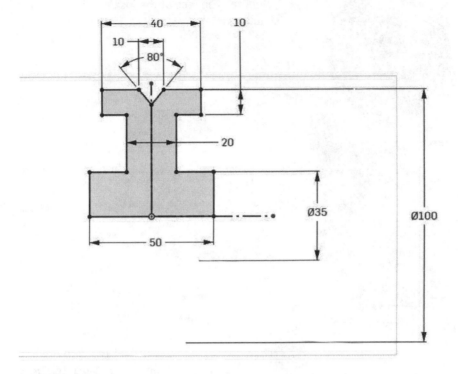

Add dimensions.
To place the diameter dimensions, select the top horizontal line, then the horizontal construction line, then place the dimension below the construction line.

12. 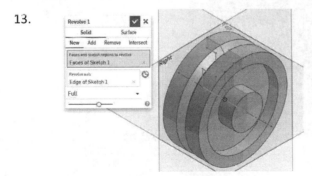 Select the **Revolve** tool.

13. Select the horizontal construction line to use as the Revolve axis.

In order for the Revolve to work, the horizontal reference/construction line must be a single line that goes through the entire sketch.

Set the Revolve to **Full**.

Green check.

14.

Select the top face of the inside cylinder for a new sketch.

15.

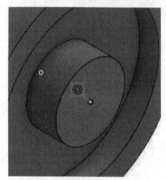

Place a concentric point on the face.

Green check to exit the sketch.

16. Select the **Hole** tool from the Features ribbon.

17.

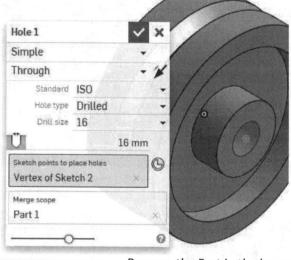

Set the Hole type to **Simple**. Set the Distance to **Through**. Select **ISO**, **Drilled**, and set the size to **16 mm**.

Select the concentric point. Green check.

18.

Rename the Part in the browser **Pulley Wheel**.

19.

Hole 1

Rename
Properties...
Assign material...
Edit appearance...
Copy...
Create Drawing of Pulley Wheel...
Export...
Hide
Isolate...
Make transparent...
Add comment
Zoom to selection
Delete...

Parts (1)
Pulley Wheel

Right click on the Pulley Wheel part.
Select **Properties**.

20.

Properties

Pulley Wheel

Name *

Pulley Wheel

Description

PULLEY WHEEL

Category

Onshape Part

Part number

1003

Revision

A

Type **PULLEY WHEEL** in the Description field.
Type **1003** in the Part number field.
Type **A** in the Revision field.

Click **Save**.

21. onshape   Close the workspace.

The Pulley Shaft

Estimated Time: 15 minutes
Objectives:
- Rename
- Revolve
- Chamfer

1.

Open the **Pulley** Workspace.

2.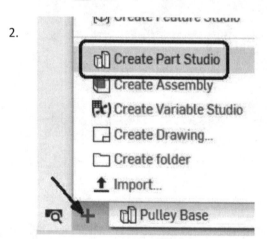

Select the + tab and left click on
Create Part Studio.

3.

Right click on the Part Studio tab and select
Rename.

Delete
Open in new browser tab
Rename
Properties...
Show code
Duplicate
Copy to clipboard
Select as document thumbnail
Move to document...

Part Studio 1

4.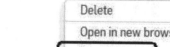

Type **Pulley Shaft**.
Green check.

5.   Select the **Front** plane for a **New sketch**.

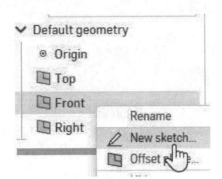

6.   Switch to a FRONT view orientation.

7. Draw the shape shown using lines.

8.

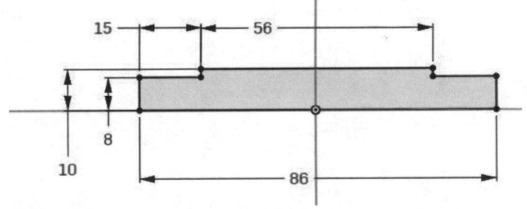

Add dimensions.

Add a midpoint constraint between the bottom horizontal line and the origin.

Add a horizontal constraint between the end points of the upper short horizontal lines.

9.   Select the **Revolve** tool from the ribbon.

10.

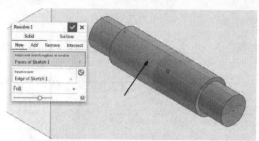

Select the bottom horizontal line as the axis.
Set the Revolve to **Full**.

Green check.

11. Select the **Chamfer** tool from the Features ribbon.

12.

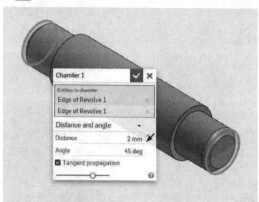

Set the Chamfer type to **Distance and angle**.
Set the Distance to **2 mm**.
Set the Angle to **45 degrees**.

Select the two outside edges.

Green check.

13.

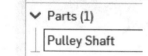

∨ Parts (1)

Pulley Shaft

Rename the Part in the browser **Pulley Shaft**.

14.

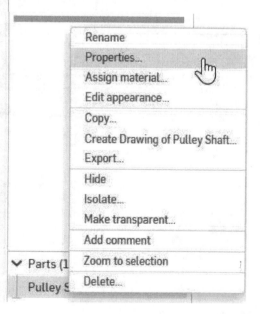

Right click on the Pulley Shaft part.
Select **Properties**.

Rename
Properties...
Assign material...
Edit appearance...
Copy...
Create Drawing of Pulley Shaft...
Export...
Hide
Isolate...
Make transparent...
Add comment
Zoom to selection
Delete...

∨ Parts (1
Pulley S

15.

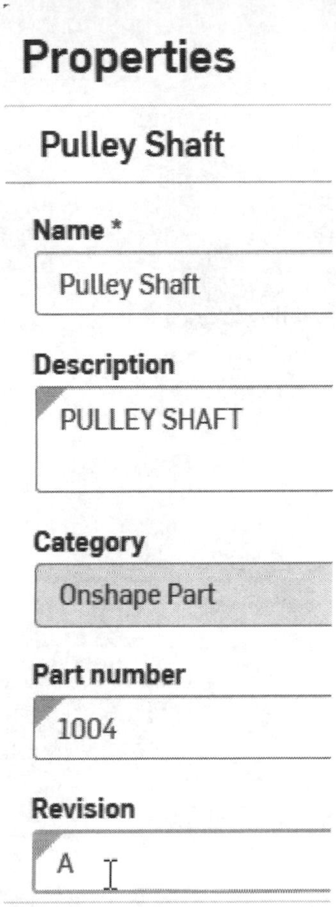

Properties

Pulley Shaft

Name *

| Pulley Shaft |

Description

| PULLEY SHAFT |

Category

| Onshape Part |

Part number

| 1004 |

Revision

| A |

Type **PULLEY SHAFT** in the Description field.
Type **1004** in the Part number field.
Type **A** in the Revision field.

Click **Save**.

16.  ⬡ onshape   Close the workspace.

The Pulley Pin

Estimated Time: 15 minutes
Objectives:

* Rename
* Extrude

1.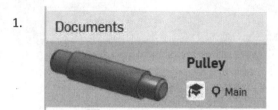

   Open the **Pulley** Workspace.

2.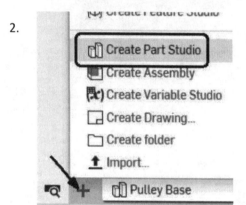

   Select the + tab and left click on **Create Part Studio.**

3.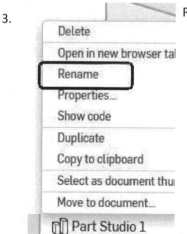

   Right click on the Part Studio tab and select **Rename.**

4. Type **Pulley Pin**.

5. 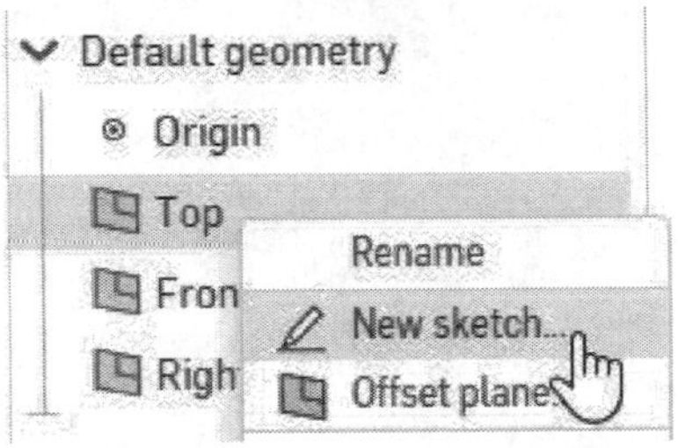 Select the **Top** plane for a **New sketch**.

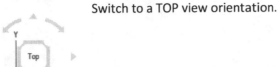

6. Switch to a TOP view orientation.

7. 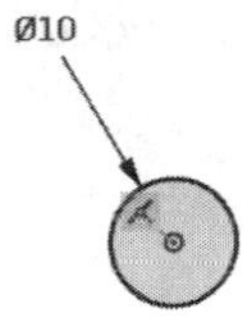 Draw a **10 mm** center circle coincident with the origin.

Add the dimension.

8. Select the **Extrude** tool.

9.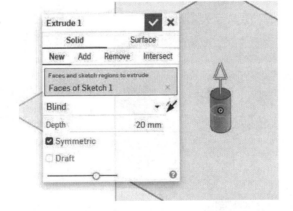

Enable **Symmetric**.

Set the Depth to **20 mm.**

Green check.

10. Rename the Part in the browser **Pulley Pin**.

∨ Parts (1)

Pulley Pin

11.

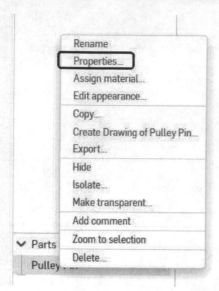

Right click on the Pulley Shaft part.
Select **Properties**.

12.

Properties

Pulley Pin

Name *

Pulley Pin

Description

PULLEY PIN

Category

Onshape Part

Part number

1005

Revision

A

Type **PULLEY PIN** in the Description field.
Type **1005** in the Part number field.
Type **A** in the Revision field.

Click **Save**.

13.

🄾 onshape

Close the workspace.

The Pulley Assembly

Estimated Time: 15 minutes

Objectives:

- Insert
- Cylindrical Mate
- Fastened Mate
- Planar Mate
- Replicate

1. 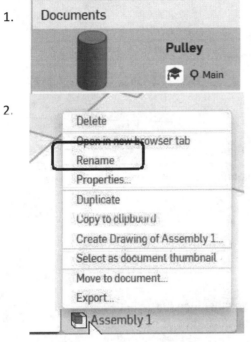 Open the **Pulley** Workspace.

2. Locate the Assembly1 tab.
   Right click and select **Rename.**

3.  Type **Pulley**.

   Green check.

4.  Open the Pulley assembly tab.
   Select **Insert.**

5.

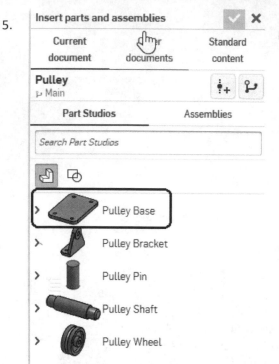

Left click on the **Pulley Base** to place first.

6.

Place two instances of the **Pulley Bracket**.

7.

Place one instance of the **Pulley Shaft**.
Place one instance of the **Pulley Pin**.
Place one instance of the **Pulley Wheel.**
Green check to close the Insert dialog.

8.

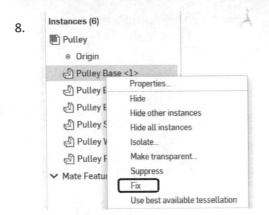

Pulley Base <1>

Highlight the Pulley Base in the browser.
Right click and select **Fix.**

9.

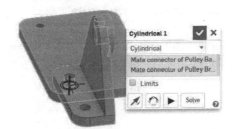

Select the **Cylindrical** mate.

Select the top of one hole in the base.
Select the bottom of one hole on the bracket.
Green check.

10.

Select the **Cylindrical** mate.

Select the top of the second hole in the base.
Select the bottom of the second hole on the bracket.
Green check.

11.

Select **Planar** mate.
Select the top of the base.
Select the bottom of the bracket.
Green check.

This will fully constrain the bracket.

12.

Repeat for the second bracket.

13.

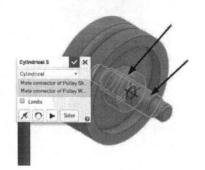

Select the **Cylindrical** mate.

Select the cylindrical face of the wheel.
Select the cylindrical face of the shaft.
Green check.

14.

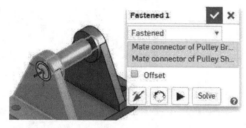

Select the **Fastened** mate.

Select the outside edge of the hole on the bracket.
Select the outside edge of the shaft.
Green check.

15.

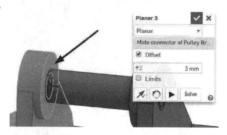

Select **Planar** mate.
Select the inside face of the bracket.

Select the top face of the wheel of the base.

16.

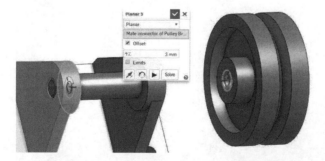

Enable **Offset**.
Set the Distance to **3 mm**.

Green check.

17.

Select the **Fastened** mate.

Select the outside edge of the hole on the bracket.
Select the outside edge of the pin.
Green check.

18.

Select the **Replicate** tool on the Assembly ribbon.
Select the **Pulley Pin**.
Enable **Match edges on face**.
Select the top face of each bracket.
Green check.

19.

The completed assembly.

Extra:

Create drawings for all parts and the assembly.

Can you locate a pin in the Standard Parts to replace the pin you created? If so, can you use the REPLACE INSTANCE tool to replace the pins?

Create an exploded view of the assembly.

Create an animation of the pulley using one of the apps in the App Store.

Create a rendering of the pulley using one of the apps available in the App Store.

Appendix: Onshape's App Store

Onshape offers a marketplace where third parties can develop and distribute applications directly to Onshape users. Each application is developed and supported by independent software vendors. Like other App Stores, each application can be rated and commented on by users. Onshape encourages their partner developers to embrace a true "freemium" model while allowing a paid version which provides more features or privacy.

Onshape wants to provide users with the greatest selection of apps possible and allow each user to try – and buy – whichever app works best for the individual user. The cost of the apps run from free to a per usage charge to a license charge. All of the paid apps provide at least one trial or trial period so you can test the app before you pay any money.

The App Store features three styles of applications:

- **Integrated Cloud Apps:** Cloud apps that run in their own Tab directly inside Onshape.
- **Connected Cloud Apps:** Cloud apps that run in their own separate browser windows.
- **Connected Desktop Apps:** Traditional installed software apps that run on Windows and/or Mac OSX.

The integrated Cloud Apps automatically update, just like Onshape. Simply refresh the browser or sign out and sign back in to load the update.

The Connected Cloud and Connected Desktop Apps run outside the Onshape browser window, but automatically load the CAD models using the Onshape API so that the most recent model is being used.

Onshape does not provide any technical support or help for any of the provided apps. If any of the apps are not working properly or you need help, you need to contact the product support for that app. Go to the App Store. Select the App. Sign in. At the bottom of the page, there is a link for Product support. Click on the link and fill in the necessary form. Be sure to share your Onshape document with the support team, so they can troubleshoot the problem.

New apps are constantly being added to the App Store, so you want to make a habit of regularly checking the App Store for new tools to boost your productivity.

I have deleted examples of how to use the apps from the text because the apps change so often it makes it difficult for the lessons to be useful. I recommend that you explore the app store and play around with different apps.

Try a Rendering and/or a Simulation app to see how you like them. Onshape is introducing simulation and rendering into the main product, but those features are currently in beta, so I won't be adding tutorials for those features until they are out of beta.

About the Author

Autodesk
Certified Instructor

Elise Moss has worked for the past thirty years as a mechanical designer in Silicon Valley, primarily creating sheet metal designs. She has written articles for Autodesk's Toplines magazine, engineering.com, AUGI's PaperSpace, DigitalCAD.com and Tenlinks.com. She is President of Moss Designs, creating custom applications and designs for corporate clients. She has taught CAD classes at Laney College, DeAnza College, Silicon Valley College, and for Autodesk resellers. Autodesk has named her as a Faculty of Distinction for the curriculum she has developed for Autodesk products and she is a Certified Autodesk Instructor. She holds a baccalaureate degree in mechanical engineering from San Jose State.

She is married with two sons. Her older son, Benjamin, is an electrical engineer. Her younger son, Daniel, works with AutoCAD Architecture in the construction industry. Her husband, Ari, has a distinguished career in software development.

Elise is a third-generation engineer. Her father, Robert Moss, was a metallurgical engineer in the aerospace industry. Her grandfather, Solomon Kupperman, was a civil engineer for the City of Chicago.

She can be contacted via email at elise_moss@mossdesigns.com.

More information about the author and her work can be found on her website at www.mossdesigns.com.

Other books by Elise Moss
AutoCAD 2023 Fundamentals
Revit 2023 Certification Guide

Notes: